TOYOTA
PERFORM[...]

A Practical Owner's Guide for Everyday Maintenance, Upgrades and Performance Modifications. Covers 1985–2005, All Makes and Models

Terry E. Heick

HPBOOKS

HPBooks
Published by the Penguin Group
Penguin Group (USA) Inc.
375 Hudson Street, New York, New York 10014, USA
Penguin Group (Canada), 90 Eglinton Avenue East, Suite 700, Toronto, Ontario M4P 2Y3, Canada
(a division of Pearson Penguin Canada Inc.)
Penguin Books Ltd., 80 Strand, London WC2R 0RL, England
Penguin Group Ireland, 25 St. Stephen's Green, Dublin 2, Ireland (a division of Penguin Books Ltd.)
Penguin Group (Australia), 250 Camberwell Road, Camberwell, Victoria 3124, Australia
(a division of Pearson Australia Group Pty. Ltd.)
Penguin Books India Pvt. Ltd., 11 Community Centre, Panchsheel Park, New Delhi—110 017, India
Penguin Group (NZ), 67 Apollo Drive, Rosedale, Auckland 0632, New Zealand
(a division of Pearson New Zealand Ltd.)
Penguin Books (South Africa) (Pty.) Ltd., 24 Sturdee Avenue, Rosebank, Johannesburg 2196, South Africa

Penguin Books Ltd., Registered Offices: 80 Strand, London WC2R 0RL, England

While the author has made every effort to provide accurate telephone numbers, Internet addresses and other contact information at the time of publication, neither the publisher nor the author assumes any responsibility for errors, or for changes that occur after publication. Further, the publisher does not have any control over and does not assume any responsibility for author or third-party websites or their content.

TOYOTA MR2 PERFORMANCE

Copyright © 2009 by Terrell Heick
Cover design by Bird Studios
Front cover photo courtesy Jeff Fazio
Interior photos by author unless otherwise noted

All rights reserved. No part of this book may be reproduced, scanned, or distributed in any printed or electronic form without permission. Please do not participate in or encourage piracy of copyrighted materials in violation of the author's rights. Purchase only authorized editions.
HPBooks is a trademark of Penguin Group (USA) Inc.

First edition: November 2009

ISBN: 978-1-55788-553-1

PRINTED IN THE UNITED STATES OF AMERICA

10 9 8 7 6 5 4 3

NOTICE: The information in this book is true and complete to the best of our knowledge. All recommendations on parts and procedures are made without any guarantees on the part of the author or the publisher. Tampering with, altering, modifying or removing any emissions-control device is a violation of federal law. Author and publisher disclaim all liability incurred in connection with the use of this information. We recognize that some words, engine names, model names and designations mentioned in this book are the property of the trademark holder and are used for identification purposes only. This is not an official publication.

CONTENTS

Acknowledgments	iv
Introduction	v
Chapter 1 MR2 Design and Development	1
Chapter 2 Building an OEM Healthy MR2	19
Chapter 3 Basic MR2 Performance Modifications & Theory	32
Chapter 4 Advanced MR2 Performance Modifications & Theory	58
Chapter 5 AW11 & ZZW30 Staged Buildups	93
Chapter 6 SW2X Staged Buildups	118
Chapter 7 Autocrossing Your MR2	167
Chapter 8 An MR2 Buyer's Guide	183
MR2 Resources	195

ACKNOWLEDGMENTS

I have gotten to know a number of wonderful, intelligent, talented MR2 people over the years, and many of them have contributed to this book. Without their knowledge, generosity and sharing spirit, this book would not have been as good. Many thanks to them, and to all others I've met and learned from along the way. Specifically, I'd like to thank: Ken Blake, Aaron Bunch, Randy Chase, "Magic" Dave, Jeff Fazio, Stephen Gunter, Tommy Guttman, David Hawkins, Bryan Heitkotter, Brian Hill, David Hillman, Garrett Katzenstein, Jensen Lum, Sarah Mays, Bill Merton, Bryan Moore, Trevor Nelson, Randy Noll, Kris Osheim and KO Racing, Bill Roberts and AutoLab, Tom Ross Jr., Bill Sherman, Bill Strong, Osman Ullah, David Vespremi, Mike Winebrenner and Chrys Zouras.

A very special thank you to Dave Martin for contributing to the very detailed, user-friendly install directions, as noted where applicable—his everyman approach to complicated bits of work will hopefully serve MR2 owners for many years; to Brian Hill and www.mr2.com; to Jensen Lum and the varied photos of his SW20; Stephen Gunter, Jeff Fazio and Bryan Heitkotter for their friendship and conversation; and to David Vespremi not only for photos of his SW20, but for allowing me to work with him on his own book, *Car Hacks & Mods For Dummies*.

Finally, a special thanks to David Goodman, who shares my passion for MR2s, can't help that he likes them in the wrong color, and who has spent a lot of time under more than one MR2 with me with a transmission on his chest.

MR2 Community—This book is the result of over fifteen years of busy MR2 ownership, and along the way I've met scores of wonderful people. Through these folks my MR2 ownership experience has been immeasurably enhanced. From the States to the U.K., New Zealand to Japan, the MR2 community is a consistently passionate and particularly communicative bunch.

Other Support

As for the non-car people that helped make this book possible, first and foremost I have to thank my father. He passed on a love and respect for the automobile that has literally changed my life, and will forever be a part of me, not to mention an unshakable will inside of me to accomplish things.

I also would like to thank my wife, Andrea, who is more wonderful than she'll ever know. It is her support and faith that pushes me every day to be better than I was the day before.

Just as critical to my productivity on a daily basis is Serena. Without you, this book would not have been possible. Welcome to the family.

I would also like to thank my daughter, Madison, you're forever my sweet pea, and Tyler and Terrell as well—you all three are my heart and soul.

INTRODUCTION

When presented with the opportunity to write this book, I was ecstatic. I've driven, cursed at, bought, sold, resented, coveted and coddled MR2s for more than fifteen years.

I have also used MR2s in about every way you can use one, from drag racing to autocrossing, hot-lapping on a road course to cross-country grand touring, commuting and showing, waxing and wrenching—and buying and selling eight (and counting) MR2s in between. In outlining this book, my main concern was not what to write, but where to begin.

What would I want in a book on the Toyota MR2? A general history? Performance recommendations? Ownership advice? A buyer's guide? These were easy to settle on. But more specifically, who was my audience, and what would they need to know? This book was written with the widest-possible audience in mind, designed to serve the most common needs of the average MR2 owner, while encouraging this "average" owner to consider their MR2 in a new light, with a focus on performance, and extracting the true sports car spirit Toyota engineered into each MR2. This was accomplished with careful attention to:

1. Planning—thus the staged approach to tuning that provides a holistic look at the MR2 as a platform for broad performance rather than simply a parts hangar.

2. Minimalist, keep-it-simple performance parts and modification schedule—if a stock part is not a restriction of some kind, generally it's left alone.

3. An "OEM healthy" MR2

4. Performance theory—thus chapters 3 and 4 explaining the reasoning behind basic and intermediate modifications.

5. Consistently advising each owner to honestly consider their goals for their MR2, rather than what everyone else is doing.

6. A wide variety of MR2 images captured in a variety of authentic, everyday ways, in addition to more formal MR2 portraiture.

The decision to focus on the broadest MR2 audience possible not only allows the largest number of owners a chance to experience their MR2 in exciting ways, but also ensures this book a place on every MR2 fan's bookshelf.

—*Terry Heick*

Chapter 1
MR2 Design and Development

This 1995 MR2 is a normally aspirated (5S-FE) North American model, and thus an SW21. Photo courtesy Toyota Motor Sales.

TOYOTA MR2, MARKET PIONEER

When Toyota launched the MR2 in Japan in 1984, it rocked the automotive world with its production mid-engine design, sporty performance and low price. Over the next 20 years, the MR2 continued to be a technological and performance trendsetter. It was introduced into the U.S. market in 1985 on the AW11 chassis, and ended production in 2005 with the third-generation ZZW30 MR2 Spyder. During that run, the MR2 was the epitome of value, price, quality and performance—a real showcase of Toyota's performance engineering capability.

Although many improvements and changes were incorporated over its two-decade production run, the MR2 never strayed far from its roots: an affordable, two-seat, mid-engine sports car.

In many ways, the MR2 still represents one of Japan's finest exported works. That is not to say that it was their most advanced or pioneering engineering effort, nor was it considered a flagship vehicle the way the Chevrolet Corvette is viewed for the U.S., or the Porsche 911 for Germany. Faster, sleeker, more demonstrative displays of Japanese automotive excellence have been produced, notably

On the outside, the MR2 Spyder was substantially different than its forebears. Beneath the skin, though, little had changed since the MR2 was unveiled in 1984. Photo courtesy Toyota Motor Sales.

MR2 PRODUCT NAMES

For the sake of simplicity and consistency, as often as possible I will refer to a single, consistent label for each MR2, either the respective chassis code or the VIN identifier.

Thus, the normally aspirated MkI will be referred to as the AW15, the supercharged MkI as the AW16, MR2 Turbos (USDM and JDM models) as the SW20, the normally aspirated MkII MR2 (USDM) as the SW21 and the MR2 Spyder as the ZZW30. A generic label refers to contexts where vague labeling by generation is sufficient and induction-type is not relevant.

AW15: 1985–1989 MR2 normally aspirated
AW16: 1985–1989 MR2 supercharged
AW11: 1985–1989 generic first-generation label
SW20: 1991–1995 MR2 turbocharged
SW21: 1991–1995 MR2 normally aspirated
SW2x: 1991–1995 generic second-generation label
ZZW30: 2000–2005 MR2 Spyder

ENGINE CODES

Like chassis codes, engine codes are comprised of alphanumeric characters used internally to designate product, only instead of entire models, engine codes refer to a specific powerplant. Further confusing the issue, there are typically multiple generations of each engine—so how do we tell them apart? Instead of renaming each refinement of an engine, as they do a redeveloped chassis, unless the engine is radically altered—adding a turbocharger and rendering a 3S-GTE from a 3S-GE, for example—it retains its given collection of numbers and letters. There are, therefore, multiple generations of the 3S-GTE—Gen I 3S-GTE, Gen II 3S-GTE, and so on, rather than the perhaps less-confusing idea of adding more numbers and letters to identify generations of engines as Honda does (B18C1, B18C5, etc.).

USDM MR2 Engine Codes

4A-GE: 1985–1989 Toyota MR2 normally aspirated
4A-GZE: 1988–1989 Toyota MR2 supercharged
5S-FE: 1991–1995 Toyota MR2 normally aspirated
3S-GTE: 1991–1995 Toyota MR2 turbocharged
1ZZ-FE: 2000–2005 Toyota MR2 Spyder

The 4A-GZE produced more torque where there was none before, though packaging the supercharger and related intercooling plumbing in such a small mid-engine car was an exercise in spatial design by Toyota.

Datsun's celebrated Z. Although the Acura NSX was brilliant, and the Mazda Miata was just as focused and interesting, the MR2 beat each of them to market by a five years. The Toyota MR2 was forward-thinking and industry leading, even if it is rarely credited for that pioneering achievement.

Though launched about the same time as the Pontiac Fiero, the Toyota MR2 was neither a response to any one competing model, nor an answer to any one question. It was a brilliantly simple runabout that stood a bit taller than everything else in the room. Greater than the sum of its parts, and arriving to a critical U.S. market suffering from rampant mediocrity, the first-generation Toyota MR2 was athletic, unique and instantly recognizable in a crowd. Quite simply, it was the right car at the right time.

Market tastes can change as fast as oil prices, so any retrospective must be careful. The Cadillac Allante that turned heads 15 years ago might seem a mistake today. In the fall of 1984, the MR2 was so far ahead of its time that the MR2 Spyder, delivered to the world over a decade and a half after the original MR2 left Japan, required no strained effort by Toyota engineers; the blueprint was done, already in so many garages, on so many magazine covers, right there on file at Toyota. Lightweight, minimalist design, mid-engine, clever powerplant

MR2 Design and Development

The SV-3 was used to gauge public reaction for the MR2 project, though at that point it was ready for production. The production AW11 was mostly unchanged. Photo courtesy Toyota Motor Sales.

and typical Toyota thinking and durability everywhere else. Though ultimately succeeding the first-generation MR2 in terms of performance, the MR2 Spyder was more market adjustment than clean-sheet design, illustrating the magical endurance of the MR2 formula.

To really understand a model, one must view it in its native historical context. Some cars can leap ahead generations at a time to challenge more modern machinery, offering up some sort of journalistic novelty. Classic Shelby Cobra versus Dodge Viper, previous-generation Porsche 911 versus the modern Cayman, let's measure the progress—you get the idea.

It is fine to compare the generations of MR2s against each other as sporting fraternal twins. However, first we must understand what the designers had in mind when they first sat down to draw these cars, and what each ended up meaning to the sports car world.

AW11 1985–1989

The 1983 Tokyo Show was the coming out party for the MR2, only it was spelled SV-3. The SV-3 was Toyota-speak for "MR2 prototype," and was for all intents and purposes what we would come to know as our AW11, minus the awkward rear spoiler and digital instrument gauge display. The SV-3

Toyota knew to be careful with the MR2. It was an important project for them at a critical time of international growth. Photo courtesy Toyota Motor Sales.

prototype was rolled out just as Pontiac was putting the finishing touches on their Fiero (perhaps an effort Pontiac might've furthered).

Prototypes are often floated to gauge public reaction, but with the Pontiac Fiero already out of the gate—one of the little MR2's few obvious market rivals—Toyota could not afford to waste time getting their project to market. When the Fiero was introduced in 1984, it was surprisingly well reviewed by the motoring press and

TOYOTA MR2 PERFORMANCE

In 1987 Toyota began offering an optional T-top roof, shown here on this 1988 model. Photo courtesy Toyota Motor Sales.

The rear taillights were significantly revised on the pace car as well, and "panda" black upper bodywork adds to its contrast with the production car. Photo courtesy Toyota Motor Sales.

For the 1987 Long Beach Grand Prix, Toyota commissioned students from a California art school to revise their MR2 for its role as pace car. Photo courtesy Toyota Motor Sales.

Among their changes were adding fog lights and a more organic front clip reminiscent of the ST165 Toyota Celica All-Trac. Photo courtesy Toyota Motor Sales.

undoubtedly motivated Toyota to get into this emerging new market ASAP. After so many bland cars during the late '70s and early '80s, the automotive industry and enthusiast gearheads were once again ready for a good, old-fashioned sports car party, something that hadn't really happened on a large scale since the 1960s.

There was a rumor for years that the AW11 was actually an abandoned Lotus design that simply Toyota purchased and produced. Perhaps this came from the fact that Lotus was involved with the design of the Supra, or it could be that the AW11 does, in fact, appear quite Lotus-like in its dimension and execution. Either way, it is a thought that refuses to die, and remains quite popular among the Toyotoa conspiracy theorists and wishful thinkers out there. Lotus is exotic, Toyota is Camry, right?

Reviewing all clear, available data, it becomes obvious Lotus never had direct, controlling involvement with the AW11. In his historical overview of the MR2 in *Toyota MR2: Coupes and Spyders*, Brian Long says, "The Japanese were quick to emphasize that the MR2's anti-dive, anti-lift and anti-squat geometry was their own work, a product of the Higashifuji Test & Research Center." I suppose we could break that sentence down and look more at what they didn't say than what they did, but the claim is good enough for me. While all manufactured product depends on influence and collaboration of some sort, the AW11 is clearly a Toyota product through and through.

Design Concerns

Finishing design issues readying the SV-3 for production centered not around horsepower, cooling, quality control, or even ergonomics,

MR2 Design and Development

The (DC2) 1997 Acura Integra Type R has numerous aerodynamic enhancements to increase downforce at speed, like the MR2, and an engine out over the front wheels that adds to a stable highway feel, unlike an MR2.

A 1985 USDM (United States Domestic Market) MR2. Note the lack of front lip, non key-colored bodywork, triangle wheels and lack of rear spoiler. Photo courtesy Toyota Motor Sales.

common sticking points for first-year production models built from clean-sheet designs. Instead, with such a short wheelbase, short overall design, and no heavy lump of metal holding down the front, there were issues with at-speed, straightline stability, something the AW11 never completely conquered even as the SW2x was introduced six years later.

"European-specification cars must achieve speeds exceeding 125 miles per hour," Akio Yoshida, in charge of the MR2 project at Toyota, is also quoted in Long's book. "To ensure safe handling at such speeds is absolutely essential, but this could not be achieved by the use of heavy steering. So we decided to use the high-caster, short-trail geometry to keep the center of gravity as low as possible to emphasize the aerodynamic performance and particularly to reduce lift." For the U.S. market, the AW11 had a rear spoiler added as standard equipment in 1986, and its front lower spoiler (lip) was modified in 1987. MR2s do tend to wander at speed, and Toyota continued to fiddle with the aerodynamics of the SW2x even as production was drawing to a close, proving that if you start something, you might as well try to finish it.

Also a concern until late in development of the preproduction models was the rear suspension design—something again they continued tweaking during its production run. Yoshida continued, "We wanted to ensure cornering at the very limit of the car's capability. In a mid-engine car, excellent cornering is almost inherent. But if the limit is high, the reaction tends to become proportionally more fierce when the limit is exceeded." Incidentally, this ultimately is the most significant handling difference between the AW11 and SW2x, with the SW2x's increased limits ensuring the opportunity for big-boy helping of ferocity when it does let go.

The subsequent suspension revisions by engineers during the AW11's run, including removal of the rear sway bar, reflected their lingering concerns over the handling character and stability, which, while detestable to automotive purists, is certainly understandable considering its focus as a mass-appeal sports car driven by owners of widely varying skill level. Factor in diverse climates, and it makes further sense. MR2s handle differently on wet roads than they do on dry.

Note: Manufacturers often make changes mid-year, which is why some 1986 MR2s got a rear sway bar, or may have either the 200mm or 212mm clutch. Oftentimes manufacturers utilize available supplies before applying new parts, and superseding old part numbers in dealership databases.

AW11 Year-to-Year Changes
1985: U.S. introduction
1986
- Rear sway bar removed
- Clutch enlarged from 200mm to 212mm
- Opt. 4-speed automatic transmission
- Opt. leather seats
- Color-keyed bodywork added, save side-view mirrors
- Third brake-light added
- Side skirts added
- Reflective clear bodywork added above engine-lid reading "Toyota"
- Rear spoiler now standard

Toyota MR2 Performance

Among the interior changes were revised door panels, and moving the parking brake handle from the driver side to the passenger. Also note the original MR2 floor mats on the 1987 model. Photo courtesy Bryan Heitkotter.

The revised rear taillights can help identify the 1987–1989 models. Photo courtesy Bryan Heitkotter.

The interior of the later, revised AW11s was more visually pleasing, but the seats suffered through the revision, more Corolla than sports car.

1987
- Engine block increased from 3 to 7 rib, smaller wrist pins and crankshaft still in production
- Stronger C-52 transmission replaces C-50
- Brakes enlarged
- Air-filter system revised (intake pathway and materials)
- Front strut-tower bar added
- Revised rear suspension geometry to reduce oversteer
- Interior significantly revised (door panels, parking brake placement, added rear speakers, seats revised)
- T-top roof added as option
- Revised rear taillights
- Front (lip) spoiler added
- Stock "triangle" wheels updated market-dependent, but often with wheels also used on (AE82) Toyota Corolla FX16 GT-S

Note: 1987 was a year of transition for the AW11 revisions, as many changes occurred midyear, with only later 1987 models receiving the revisions.

Note: 1987–1989 models are more receptive to a 4A-GZE (supercharged) engine swap than the early 1985–1986 models.

1988
- Revised ECU
- Slightly less aggressive camshafts
- Revised fuel injectors, now larger and high-impedance
- Supercharged model (AW16) added w/4A-GZE engine. This model also included the following:
  - E-51 transmission to replace the C-52 (stronger, slightly more relaxed gearing to take advantage of additional 4A-GZE torque)
  - Raised engine-lid cover
  - "Supercharged" decals
  - Teardrop wheel design
  - Revised instrument cluster to include "supercharger on" light
  - T-top roof standard

1989
- Brake master cylinder enlarged
- 10mm rear sway bar

Even with the slightly increased power (112 hp to 115), the later AW15s lost some of the edge the early cars had, certainly no quicker in a straight line, and far more prone to stubborn understeer during at-limit handling. Overall, this revision rendered a gentler MR2, but there were no significant performance enhancements other than a stronger transmission.

MR2 DESIGN AND DEVELOPMENT

Note the "SUPERCHARGER" script on the passenger door and the hardtop which is not present on USDM AW16 models—this is a JDM model. Photo courtesy Kenzo Nagai.

Something old, something new: 1989 AW15 with ZZW30 wheels. Cheap, lightweight wheels with OEM quality, and still all MR2.

4A-GE: TOYOTA'S CROWN JEWEL

The 4A-GE is a flexible workhorse that has been revised continuously by Toyota since its inception in the early 1980s to fit a variety of applications, from mid-engine sports car to economical coupe. Toyota must be a sentimental manufacturer, sticking for so long with the same basic engine design. It also is rather fuel efficient for such a single-minded little engine, with healthy, carefully pedaled versions capable of 32+ MPG in highway driving, and very little maintenance required other than what is factory-recommended.

Over the course of its life the 4A-GE saw it all: TVIS, large intake ports, small intake ports, supercharging, 20-valve heads, VVT—everything but a turbocharger (something HKS first remedied, and others have since followed suit). It also saw duty in front-drive models, rear-drive coupes and hatches, and one fantastic mid-engine sports car as well. Every year, like clockwork, some version of 4A-GE could be expected in some model, in some market in the world. Other long-serving, flexible "pet" powerplants from major manufacturers exist, among them the SR20 from Nissan, the B6 from Mazda, the 4G63 from Mitsubishi, and Honda's stalwart B16 and B18, but none has been as thoroughly wrung-out, redesigned, stretched, and reworked as the 4A-GE. It is a legendary engine in its own right, perhaps even more so than any of the platforms it has powered.

The 4A-GE was among the earliest mass-production, inline, 4-cylinder, dual-overhead camshaft, 16-valve engines. The "G" in 4A-GE designates a Toyota wide-angle performance head, with a 50-degree valve angle, and superior flow in and out of the head. The "G" head is opposed to

Replacing the factory steering wheel on the AW11 is less of a sacrifice in safety than it is on later MR2s, considering steering wheel air bags did not come along until the SW2x—1991 in the U.S.

the narrow-angle "F" heads on other engines—the 3S-FE for example, or the 4A-FE, which was the 4A-GE's more frugal counterpart. From the beginning it was meant to display Toyota's engineering ability and love for performance, an intent that has certainly waned in the early 21st century.

The 4A-GE features a non-interference head, meaning that should the timing belt break, the pistons and the valves should not contact (and destroy) one another. It uses a shim-over-bucket

The 3-rib 4A-GEs had lighter engine components, making them more willing to rev, and offering a satisfying driving experience even with less power output than later versions.

design as well, so that the valves can be adjusted without removing the camshafts, certainly a thoughtful design element. The intake manifold was equipped with a technology called TVIS—Toyota Variable Intake System. This system was Toyota's way of enhancing the low rpm performance of such a small displacement engine—at only 1587ccs, and normally aspirated to boot, they had their work cut out for them. The TVIS system was basically an 8-runner intake manifold that had half of the runners blocked by a vacuum-operated butterfly valve until a certain rpm was achieved—depending on load usually around 4350 rpm on the 4A-GE. This system would later be used on the MR2 Turbo's 3S-GTE as well.

Blocking the runners increased torque at lower rpm, but still allowed for the full airflow capability of the manifold to be used as rpm increased. It sounds simple, and possibly underwhelming, but judging it by today's standards isn't fair, of course. This Jekyll/Hyde concept is the basic premise behind the variable valve lift and timing engines that have become commonplace today, fitting an engine with two (or more) personalities, thus allowing for more flexible and ultimately more potent engine setups, with less overall compromise in design and performance. Honda may get all the press for progressing engine design, but Toyota gives up little to anyone in that regard. The iron block, aluminum head and relatively simple overall implementation of the 4A-GE makes it extremely durable, even in its early 3-rib form capable of supporting around 200 horsepower given the proper modifications. Again, this may not sound like much today, where Miata engines can shoulder 300+ hp all day long, but remember that the 4A-GE was busy cutting its teeth in the very early 1980s, when Michael Jackson was still the coolest person on the entire planet. The 4A-GE features a strong iron block, and especially sturdy crank and connecting rods, overall a very robust design for its vintage and overall production cost.

Toyota's Revision of the 4A-GE—Sometime in 1987 the later 7-rib block was introduced, and was even more durable, with those 4 additional "ribs" added to the block, oil-squirters on the pistons, and even adjustments to the metallurgical content of the block to increase its strength. While the 7-rib large-port was indeed sturdier, many purists criticized the now-heavier internals of the 4A-GE, all of that extra strength bringing with it a heavier reciprocating mass, somewhat harming the 4A-GE's eagerness to rev, undoubtedly its best characteristic. Still, the engine was more or less the same 4A-GE everyone had grown to love, and its new design could now foster even more horsepower in modified form due to that added strength.

Somewhat further harming the rev-happy nature was the supercharger Toyota would make an option in the 1988 MR2. This sacrifice, though, was not without considerable gain. The 4A-GE—or the 4A-GZE actually—now had torque, and was extremely quick even in stock form. In fact, in 1988 and 1989 one of the quickest cars in the world from 0–30 was an automatic transmission-equipped supercharged MR2, answering one of the biggest criticisms of the 4A-GE. Still, the 4A-GE was on borrowed time, if for no other reason due to the meager 1.6L engine displacement. The 4A-GE was starting to grow a little long in the tooth, and its supercharging gave the powerplant some added time and application.

While initial development of the 4A-GE rendered stronger internals and a supercharged variant, Toyota next fiddled with head design, coming up with the small-port 4A-GE. This engine sported the robust block from the supercharged 4A-GZE—the 7-rib wunderblock mentioned above—and smaller intake ports in the head for increased intake velocity, negating the need for TVIS. This even further simplified the 4A-GE's function, and with an increase in compression from 9.4 to 10.3, increased horsepower to a Toyota-claimed 140 horsepower, though that figure might be a bit optimistic. While some low rpm performance was sacrificed with this design even with the small-port head and higher compression pistons, it undoubtedly offered stronger performance overall than the large port, coming alive above 5000 rpm. In truth, the amount of torque produced below 3000 rpm should not be a point of criticism for any normally aspirated 1.6-liter engine. When cost and complexity of installation is considered, the small port is considered by some as the best overall option for the AW11 for a pure driving experience, and is stout enough to support significant aftermarket upgrades.

Others might indeed prefer the 4A-GZE, certainly not a wrong-headed thought. Seeing duty in the variations of Corollas in the Japanese market, this supercharged engine would continue to see enhancements over time as well, among these changes additional boost, a MAP-sensor setup, an increase in compression, and other design changes such as a small-port head that dovetailed behind the normally aspirated engine's enhancements. When importing these engines for your MR2, stick to the cleanest 4A-GZE you can find, regardless of its revision. The fact that they are supercharged allows horsepower increases to be had far more easily than normally aspirated versions, making the difference between them less critical than the gap between 16 and 20-valve 4A-GEs.

In 1991 Toyota again began a redesign and production of a further revised 4A-GE, revising the 4A-GE obviously Toyota's favorite pastime. This round of changes included lighter internals compared to the robust late 7-rib blocks and small-port engines, a tubular exhaust manifold, fantastic individual throttle-bodies, and a 20-valve head (3 intake, 2 exhaust) that also featured variable valve timing (VVT). This busy engine, known as the "silver top" 20-valve 4A-GE due to its silver valve cover, still featured an iron block, continuing to allow aftermarket forced-induction as a possibility, and produced 160 horsepower. Toyota would later amend this jewel of an engine as well, rendering the "blacktop" 20-valve. The blacktop featured a higher compression ratio of 11:1, again lightening the internals, and moving from an AFM setup to MAP, ultimately rewarded with another 5 hp, now up to 165, and finishing off the 4A-GE's brilliant, Cal Ripken-like run.

Which 4A-GE Is the Most Mod-Friendly?

Brilliant pioneering spirit and considerable strength aside, the 4A-GE was mostly torqueless and high-strung. While the 20-valve did indeed produce as much as 165 horsepower, it still lacked the overall muscle and tractability of the supercharged 4A-GZE. Expensive, insane normally aspirated 10000 rpm engines notwithstanding, to get 1.6 liters to produce significant horsepower requires some sort of force-feeding. Thus arranged and suitably modified, the 4A-GE is capable of somewhat reliably producing 200+ horsepower, with the block itself capable of supporting more with appropriate tuning and lots of boost. As is the

The ultimate incarnation of the 4A-GE, the legendary Formula Atlantic 4A-GE: 12.7:1 compression ratio, 240 hp. Photo courtesy Aaron Bown.

case with the 3S-GTE and any other highly tuned forced-induction engine, "proper tuning" amounts to superfluous intercooling, careful turbocharger sizing, and proper engine management. In fact, this can be said of almost any engine: they are generally as strong as their tuning. Beat it to death with detonation, and any engine will eventually fail.

For light modification under 175 hp, among the 16-valve engines the early 3-rib version works well, considering the lighter internals and somewhat more "streetable" torque. The 3-rib engine has an eager, easy feel to it that certainly has a charm, similar to the early 1.6L Miata engines versus the later 1.8L engines. The 3-rib block isn't a liability at these power levels, and the large-port head supplemented well by the TVIS intake manifold.

If you plan on going beyond 175 hp, the later 7-rib large-port and small-port designs become more suitable. The stronger block is actually put to use at these power levels, and the rpm necessary to produce these power levels will really put a small-port head to effective use. For even bigger power, an ideal candidate is a 4A-GZE with the supercharger removed, and a turbo in its place, something the 4A-GZE internals should support well with proper tuning, provided boost levels stay below 12 psi. The 7-rib block, small-port head, oil squirters, and larger connecting rods will be more durable to better handle the rod-pounding torque of forced-induction. Keep in mind though, regardless of relative internal engine strength, any 4A-GE/4A-GZE that is going to see more than 10 psi of boost will really need thorough EMS tuning to thrive.

The 20-valve 4A-GE will offer lots of nice gadgets

4A-GE DIFFERENCES

Name:	3-rib large-port	7-rib large-port	Small-port	Silvertop 20-valve	Blacktop 20-valves
Production Run:	1984–1987	1987–1989	1989–1993	1992–1996	1996–1999
Displacement:	All 1587cc/1.6-liters				
Induction:	N/A	Normal	Normal	Normal	Normal
TVIS:	Yes	Yes	No	No	No
Cam and Valve:	DOHC/16v	DOHC/16v	DOHC 16v	DOHC 20v	DOHC 20-valve
Intake Port Size:	Large	Large	Small	ITB	ITB
Piston Pin:	18mm	20mm	20mm	20mm	20mm
Connecting Rod:	40mm	42mm	42mm	42mm	42mm
Air Metering:	AFM	AFM/varies	MAP/varies	AFM	MAP
Block Ribs:	3	7	7	7	7
Oil-cooled Pistons:	No	No	Yes	Yes	Yes
Horsepower (varies):	112 @ 6600 rpm	115 @ 6600 rpm	140 @ 7200 rpm	160 @ 7400 rpm	165 @ 7800 rpm
Bore x Stroke:	81mm x 77mm	81mm x 77mm	81mm x 77mm	81mm x 77mm	81mm x 77mm
Compression:	9.4	9.4	10.3	10.5	11.1
Injector Size:	182cc, Top	182cc, Top	235cc, Top	295cc, Side	295cc, Side
Octane:	Standard	Standard	Premium	Premium	Premium
Visual:	# of ribs on block	# of ribs on block	Ribs on back of head	ITBs, valve cover color	ITBs, valve cover color

The terms "blue top," "red top," and other methods of identifying 4A-GEs by colors on their valve cover is inconsistent. Refer to the above row "Visual" to identify 4A-GEs.

The introduction of the 7-rib large-port reduced the power in Japanese models from 130 hp to 120hp (emission modifications), while output was increased in the American market from 112 to 115.

Air metering equipment, emissions equipment, octane requirement, and consequently horsepower vary by market.

The practice of measuring and reporting engine output (in torque and horsepower) has changed since 1985—and also, as with so many other things, varies by market.

ITB signifies "Individual Throttle body," a 4-throttle/throttle body setup on the 4-cylinder 4A-GE.

right off the bat, and if 160-165 hp is all you want, and you want it normally aspirated, this is a strong choice, preferable to modifying the 4A-GE in many cases. They can, however, get expensive in regards to time and money if you want to modify it further, or run into issues sorting problems.

The strength of the 4A-GE lies in its head design, so flow there is rarely an issue, though if you are building an engine to see high rpms, above 7500 rpm, head work would indeed pay dividends. At the risk of sounding repetitive, though there are indeed differences in bottom-end strength across the range of 4A-GEs, the real key to relatively safe, significant horsepower gains is tuning. If you want significant horsepower in a 4A-GE, start with a strong 7-rib 4A-GZE block, choose your method of power-adding—supercharging, turbocharging, revs, or nitrous—and spend your money in EFI/EMS tuning.

SW2X 1991–1995 TOYOTA MR2

The SW2x was a simultaneous effort in revision and progression by Toyota, a blended approach of

Part of what made the 20-valve so fantastic were the motorcycle-like individual throttle bodies. They also up the complexity factor of proper modification and underscore the need for appropriate air/fuel tuning. Photo courtesy Chrys Zouras.

AW11 ENGINE SPECIFICATIONS
4A-GE 3-rib, Large-Port, AFM
Block/Head Type: Iron/Aluminum alloy
Displacement: 1.6 liters (1587cc)
Bore & Stroke: 81mm x 77mm
Compression Ratio: 9.4:1
Cam and valve arrangement: DOHC, 16 valves
Maximum Horsepower: 112 hp @ 6600 rpm
Maximum Torque: 97 lbs-ft. @ 4800 rpm
Redline: 7500 rpm
Induction: Normal
Recommended Fuel: Unleaded Standard
Fuel capacity, gallons: 10.8
Oil capacity, quarts: 4.4
Horsepower per liter: 70 hp per liter

Revisions to the 4A-GE in 1987 allowed the 4A-GE to produce 115 hp, and 100 lbs-ft. of torque.

4A-GZE Large-Port, AFM
Block/Head Type: Iron/Aluminum alloy
Displacement: 1.6 liters (1587cc)
Bore & Stroke: 81 x 77mm
Compression Ratio: 8.0:1
Cam and valve arrangement: DOHC, 16 valves
Maximum Horsepower: 145 hp @ 6400 rpm
Maximum Torque: 140 lbs-ft. @ 4000 rpm
Redline: 7500 rpm
Induction: Supercharged
Recommended fuel: Unleaded Premium
Fuel capacity, gallons: 10.8
Oil capacity, quarts: 4.4
Horsepower per liter: 90.63 hp per liter

Later non-AFM, MAP-versions of the 4A-GZE, which never appeared in the AW16, made up to 170 hp, mostly via increases in compression to 8.9:1, and higher boost levels.

The 1991 MR2 Turbo was a boon to Toyota's starchy image but was soon roasted by the press as a "handful." Those tiny rear tires didn't help. Photo courtesy Toyota Motor Sales.

Most 1991–1993 SW20s were T-tops, while the others were the less common sunroof turbo models, and finally the rare hardtop turbo. Hardtop 1994–1995 SW20s were not available. The sunroof offered extra structural rigidity.

bolder engineering and brazen styling. The AW11 had seen some enhancements during its pioneering run, but the SW2x was a very different beast altogether. Both were mid-engine, rear-drive, two-seat Toyotas, but the similarities ended there. Owing to the time period in which they were designed, the SW2x was aesthetically much smoother than the AW11. While any sports car should offer an exhilarating driving experience, SW2x Chief Engineer Kazutoshi Arima admitted that "style" was also among the chief targets of the SW2x; on the AW11, style was more of a byproduct of design.

Toyota also continued to adjust the aerodynamics, adding a lower front lip in 1993, and turning the rear spoiler into a one-piece affair in 1994. The Japanese market SW2x also received a new rear spoiler yet again in 1997 (its third design), and speed flaps were also developed and sold by Toyota in other markets to further manipulate the airflow over, under, and around their feisty little sports car. Toyota has continuously sought out aerodynamic enhancements for the MR2, most engineered to increase the stability of the car at speed. Despite Toyota's best efforts, though, most MR2s will never feel as nailed down as one of their longer-wheelbase, front-engine sports car counterparts.

While approximately 400 lbs. heavier than the lightest AW11, the SW21 got a dose of displacement when it received the 2.2-liter engine from the Camry, a long-stroke engine more calm than exciting. In spite of its yawning performance, this engine broadened the appeal of the SW2x, making it more affordable and less intimidating for those put off by the word "turbocharged." In all

Toyota MR2 Performance

In 1994, Toyota replaced the early 3-piece spoiler with a 1-piece spoiler that revealed some separation from the trunk at its sides. Photo courtesy Brian Hill.

You can identify this 1994 MR2 as an SW21 by the smaller-diameter exhaust tips; from this angle, the non-raised engine lid louvers cannot be seen. Photo courtesy Toyota Motor Sales.

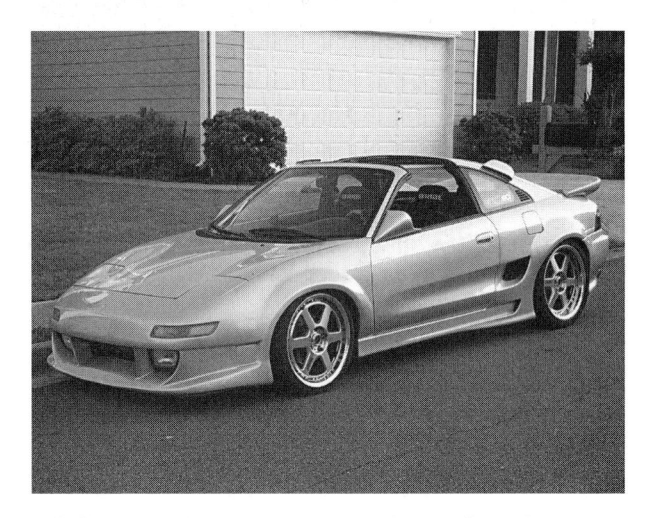

This is a 1994 MR2 Turbo, though with the front lip, steering wheel and rear spoiler all aftermarket pieces, and with no view of the revised taillights, you'd have no easy way of spotting that yourself. Photo courtesy Brian Hill.

honesty, that turbocharger can be considered a metaphor for the celebrated engineering excess that got the MR2 into deep water to begin with, ultimately responsible for the MR2's five-year hiatus from the American market in the late 1990s.

The car was simply too much: handling too sensitive, storage too small, for some too fast and perhaps most crucially, too expensive.

Of course, pick a used one up now and it seems just right, but new automobile manufacturing is all about demographics, marketing targets, product and profit. What made the MR2 so brilliant—the final, engineered product—was what ultimately failed it in the marketplace. The market window for two-seat, mid-engine turbocharged sports cars approaching $30,000 had been shut.

SW2x Year-to-Year Changes
1991: Introduced to U.S.
1992: No significant changes

1993
- Rear suspension geometry revised
- Caster made non-adjustable
- Suspension lower, stiffer
- E153 SW20 transmission strengthened, revised
- Brakes enlarged
- Throttle-body inlet now non-tapered
- Distributor cap and spark plug wires revised
- Wheels enlarged to 15", redesigned; tires widened
- Front spoiler (lip) added

1994
- Rear taillights revised
- Rear spoiler revised
- Passenger-side air bag added
- Minor interior improvements, including a smaller steering wheel and revised shift knob

1995: No significant changes

3S-GTE Engine Analysis

In addition to progressive styling and design, the brand-new SW20 featured a new engine as well. The 4A-GE lived on in other more plebian Toyota designs, but there was a need for an exciting new powerplant to further separate this new MR2 from its previous design. Toyota went with an amended version of the 3S-GTE used in the ST165 Celica All-Trac, this revised 3S-GTE producing 200 horsepower and 200 lbs-ft. of torque, enough to provide the SW20 with impressive acceleration: 0–60 in 6 seconds, and 0–100 in about 16 seconds in stock form.

Based off the 3S-GE, this second-generation 3S-GTE was a 2.0-liter engine featuring an iron block, a twin-cam 16-valve aluminum head, and a unique twin-scroll turbocharger. Featuring the same square 86mm x 86mm bore and stroke every 3S-GTE had, an 8.8:1 compression ratio, and a large-port

MR2 Design and Development

The early SW2x offers sharp—but admittedly unforgiving—handling. Wider rear wheels and tires and adjustable Tokico shocks, as equipped on this 1992 SW20, can help to reduce this tendency and are relatively simple and inexpensive to install.

Side by side, the SW2x made a much stronger visual impact and gave the appearance of a much larger—and sleeker—vehicle.

This tuned 3S-GTE, in the middle of a turbo swap, features 8.5:1 (compression ratio) JE forged pistons, stock cams and a (forthcoming) dual ball-bearing T3/T4 turbo, all generating around 300 whp at 18 psi.

cylinder head, the Gen II 3S-GTE was a capable performer stock. And perhaps more importantly, it features extremely stout internal construction, including particularly strong crankshaft and connecting rods, and hypereutectic pistons that endow it with more than sufficient strength to support significant additional horsepower.

While rather simple in design and slightly heavy compared to modern aluminum block forced-induction engines, the 3S-GTE was everything it needed to be: simple, cost-effective and robust. With appropriate external modifications the 3S-GTE can reliably make 300–400 whp on stock internals, assuming, of course, quality EMS tuning.

Continuing 3S-GTE Development—Of course, Toyota could not leave well enough alone. By 1994 the Gen II 3S-GTE, while modestly powerful, was already four production years old, and certainly could not keep up with the pending third-generation FD3S Mazda RX-7 TT, much less the big dog Nissan GT-R Skyline, Acura NSX, and Mitsubishi 3000 GT VR-4, even in 225 hp JDM form. While certainly a different car than those heavy hitters, and at a less-expensive price tag, Toyota continued to enhance the 3S-GTE, releasing the Gen III 3S-GTE to the Japanese market SW20 in 1994, and bolting it into the engine bays of the ST205 Celica All-Trac as well. By this time the SW2x was already on its way out in the USDM however, so rather than bother with emissions-certifying this new engine in the United States, Toyota continued producing the USDM MR2 Turbo (SW22) with the Gen II 3S-GTE through its USDM demise in 1995.

The 3S-GTE would see one more round of refinement over the years, in the form of a fourth-generation 3S-GTE at least used in the JDM Toyota Caldina (we cannot state with authority whether or not Toyota snuck the 3S-GTE into some uber-rare Japanese tractor—they do, after all, also manufacture forklifts). Among other changes, the Gen IV 3S-GTE featured a 9.0:1 compression ratio, and 260 hp.

Which 3S-GTE Is Best for Modifying?—Not including the Gen I 3S-GTE found in the early (ST165) Celica All-Trac, any Gen II or Gen III 3S-GTE is capable of producing all the horsepower money can buy. Featuring iron blocks and wonderful head designs, you can do whatever you need with either, though both 3S-GTEs have their

Toyota MR2 Performance

The T3/T4 from Innovative Turbo Systems is a big step up from the stock CT26 turbocharger but is a crucial modification: The turbocharger clearly makes the engine, rather than the reverse. The configuration shown here is utilizing a unique manifold adapter in an attempt to maintain the stock exhaust manifold for simplicity's sake.

The 3S-GTE in any form is a capable engine, with a strong bottom end and an efficient cylinder head. The revisions to the Gen II that produced the Gen III were focused on airflow and enhanced durability, which makes the Gen III 3S-GTE an excellent choice for a wide variety of street, road course and drag applications. Photo courtesy Vespremi/Crites/Smith.

The Gen III, through significant changes to the head and turbocharger, developed 245 hp, an increase of 45 hp over the USDM Gen II. Anything look out of place to you here? The unique intake path maybe? This is a Gen III S-GTE in an AW11. Photo courtesy Tommy Guttman.

pros and cons.

There is a debate in MR2 circles centering around the third version of the 3S-GTE produced by Toyota. As we've described, our SW20—the USDM 1991–1995 MR2 Turbo—was powered by the Gen II 3S-GTE, and produced a very symmetric 200 horsepower and 200 lbs-ft. of torque—225 horsepower in JDM form. Sadly, the U.S. market never received the Gen III 3S-GTE.

This debate really only becomes an issue for those swapping engines. The vast majority of MR2 Turbos still available have a lot of miles on them. If your 3S-GTE is on its last oil change, or your MR2 is normally aspirated and you're itching for more, an engine swap is probably your best bet. Given modest goals, really any 3S-GTE will do. As a production USDM engine, the Gen II 3S-GTE enjoys more OEM part support from local dealers, and has more immediate aftermarket support as well. If you simply want a stock, reliable 200 horsepower, the Gen II is your baby. But what if you're shooting for the stars?

If you absolutely must have the most powerful MR2 on the road—then the answer is probably still a Gen II 3S-GTE, though at that point so much of the stock configuration must be replaced that what you're left with is barely Toyota at all, much less representative of any arguable difference between Gen II and Gen III. This much is generally agreed upon by most knowledgeable MR2 tuners. Up until this point, there is no real debate.

The Gen III 3S-GTE features a boost-friendly 8.5:1 compression ratio, more potent turbocharger and 540cc injectors capable of fueling your 300 horsepower goals (versus 440cc in the Gen II). There are no more TVIS concerns and no more worries over whether your Gen II ECU is doing funny things with the timing, or whether or not your turbocharger is going to fall on its face 1500 rpm short of redline. You get a stock steel head gasket, enhanced cooling in the head, enhanced cam profiles, a bigger bore fuel rail, a bigger throttle body, superior head design, superior intake

MR2 Design and Development

This is a 1993 SW22, or North American 3S-GTE MR2 Turbo. You can just make out the non-original taillights, which are in the process of being upgraded to 1994–1995 units, and the non-original side-molding that says TWIN CAM 16 turbo. What you cannot see are the polished OEM wheels, which were available only on later JDM SW20s.

Among a crowd of mechanical enhancements, Toyota also added an extended front lip spoiler to the 1993 models. This 1994 turbo is also hiding its revised taillights. Photo courtesy Toyota Motor Sales.

manifold and a handful of other tweaks to make Gen II owners envious. The Gen III 3S-GTE removes many of the worries of aftermarket tuning out of the box, which is a good thing. Keep it simple.

ZZW30 2000–2005 TOYOTA MR2

Lessons had indeed been learned with the extremely aggressive SW20, and the subsequent MR2 Spyder is probably best viewed with this market consideration and minimalist design in mind. Though coming to the USDM five years after the SW2x had bowed out, the ZZW30 was very much a reaction to the SW2x's failure as much as it was born of more proactive focus.

If nothing else, the Spyder had to be affordable and lightweight. Of the three generations, it ended up being by far the lightest, and the probably the best handling, as a convertible no less!

Overall, the third-generation MR2 should be viewed as a success, especially considering its original "MR-S" labeling in its native Japan,

SW2X ENGINE SPECIFICATIONS

5S-FE
Block/Head Type: Iron/Aluminum alloy
Displacement: 2.2 liters (2164 cc)
Bore & Stroke: 87mm x 91mm
Compression ratio: 9.5:1
Cam and valves: DOHC 16 valves
Maximum Horsepower: 135 hp @ 5400 rpm
Maximum Torque: 145 lbs-ft. @ 4400 rpm
Redline: 6000 rpm
Induction: Normal
Recommended Fuel: Unleaded Standard
Fuel capacity, gallons: 14.3
Oil capacity, quarts: 4.4
Horsepower per liter: 65.91 hp per liter

Notes: 1991–1993 5S-FE MR2s produced only 130 hp and 140 lbs-ft. of torque.

5S-FEs meeting California emissions standards made 5 fewer hp, with the 1995 Toyota MR2 press kit quoting 130 hp @ 5400 rpm, and the same 145 lbs-ft. of torque at 4400 rpm.

3S-GTE Large-port, TVIS Intake Manifold
Block/Head Type: Iron/Aluminum alloy
Displacement: 2.0 liters (1998cc)
Bore & Stroke: 86mm x 86mm
Compression ratio: 8.8:1
Cam and valve: DOHC 16 valves
Maximum Horsepower: 200 hp @ 6000 rpm
Maximum Torque: 200 lbs-ft. @ 3200 rpm
Redline: 7250 rpm
Induction: Turbocharged
Recommended Fuel: Unleaded Premium
Fuel Capacity, gal.: 14.3
Oil Capacity, qts.: 4.4
Horsepower per liter: 100 hp per liter

Notes: The Gen I version of the 3S-GTE made 180 hp. The JDM Gen II version of the 3S-GTE, with the ceramic CT26 turbocharger, made 225 hp. This hp increase over the 200 hp USDM version came courtesy of increase in boost via the ceramic turbocharger, and more aggressive ECU settings. The Gen III 3S-GTE, with its multitude of revisions, made 245 hp, an outstanding 122.5 hp per liter. No California-emission MR2 Turbos were available in 1995.

Toyota MR2 Performance

No top, no worries—and no trunk or turbocharger either. Photo courtesy Toyota Motor Sales.

Quite the departure from the SW2x, the ZZW30 continues the tradition of each successive MR2 being some sort of a departure from the previous. Here, the front mouth of this later ZZW30 turns up, more like a smile than the early version. Photo courtesy Toyota Motor Sales.

ZZW30 taillights on an early ZZW30. Revised or unrevised, the ZZW30 was significantly different from the SW2x because automotive culture had changed significantly since the early 1990s. Photo courtesy Toyota Motor Sales.

indicating Toyota's intentional move away from the traditional MR2 label, image and approach. Over its USDM production run, revisions were fairly minor, the most significant occurring in 2002 when the SMT transmission was initially offered, and 2003, when the rear wheels grew from 15x6.5 to 16x7 to increase the existing front-to-rear stagger, perhaps in a move towards enhanced stability through increased rear grip.

Achieving 0.91g on the skidpad, and quarter-mile times of 15.6 stock in tests by *Road & Track*, clearly the ZZW30 was a decent performer bone-stock, and its most modest feature—the normally aspirated 1ZZ-FE engine—was obviously not a temptation for revision like previous MR2 powerplants. In this case, Toyota left well enough alone, and the MR2 Spyder stayed light, simple, and relatively inexpensive.

MR2 Spyder Origins

To be fair, the ZZW30 may not have ever been meant to accomplish more than it did. We can view it from a different vantage, as an MR2 purist, but the automotive economy is becoming more business and less emotion. That is not to suggest that the sky is falling in regards to sports car design; certainly they are getting faster, safer and more affordable across the board. But the reality of modern economics has affected automotive manufacturing, more so than in years past when halo cars were given much more flexibility in regards to bottom line. Cars like the JZA80 Toyota Supra may not have seen the light of day in 2009. Current corporate philosophy mandates that sports cars make money, rather than simply "get people in the showroom," and when they don't, they are removed from production swiftly, with little regret.

The ZZW30 was drawn up and assembled during a time when Toyota's "fun" image suffered considerably, and together with the Celica injected some music into Toyota showrooms across the country, though both have disappeared as quickly as they arrived. We can deride the ZZW30's available luggage space or powerplant character forever, but it mostly accomplished what Toyota intended it to. If we do, in fact, get another MR2 in the USDM, its design will undoubtedly be filtered through a similar confluence of economics, design and environmental concerns—concerns that just may indeed yield something similarly polarizing to the enthusiast community. Mass-produced, affordable, no-compromise sports cars may be an old-fashioned idea. Time will tell.

MR2 Design and Development

Later, revised ZZW30 rear taillights and rear bumper design. These taillight revisions came to the ZZW30 in its fourth year of production in the U.S. market, as was the case with the SW2x and its taillight-revision. Photo courtesy Toyota Motor Sales.

ZZW30 Year-to-Year Changes
2000: Introduction to U.S.
2001: No significant changes
2002: SMT transmission now optional
2003
- SMT transmission revised, now with 6 forward gears instead of 5
- Revised spring and shock settings
- Additional chassis/member bracing
- Rear wheels enlarged, tires widened
- Revised front-end, fog-lights standard
- Revised taillights

2004: Limited slip differential now optional
2005: 6-Disc CD changer available

The 1ZZ-FE Engine: A Modest Proposal

Of all of the MR2 powerplants, the 1ZZ-FE from the ZZW30 has to be the least appreciated. The 4A-GE, judged in proper historical context, was nothing short of brilliant. So, too, was the supercharged 4A-GZE. And the sky-is-the-limit 3S-GTE was perhaps surpassed by only Mitsubishi's 4G63 in overall success. The 1ZZ-FE never stood much of a chance, ironically very much the victim of the 3S-GTE-powered MR2's financial failure in many ways.

Though the 1ZZ-FE was very similar in terms of output to the 5S-FE, the 5S was the economical alternative to the forced-induction MR2 Turbo, so if nothing else was simply meant to provide a sure-footed option for the frugal-minded buyer. The 1ZZ-FE, though, was it for the MR2 Spyder. The beginning and the end. If you wanted an MR2 Spyder you were getting exactly 138 hp, no matter how capable the chassis, and it was normally aspirated, making additional horsepower tuning far more difficult.

Viewing the 1ZZ-FE in Hindsight—Although a perfectly acceptable engine that motivated the MR2

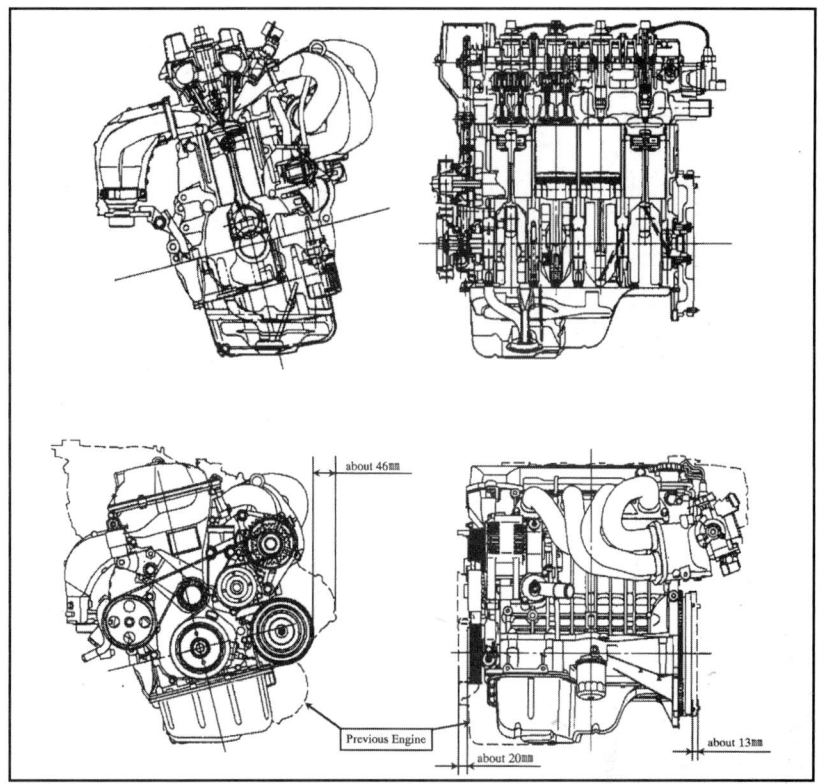

These line drawings from Toyota highlight some of the changes over previous 4-cylinder engines considered predecessors. Toyota's 1ZZ-FE does a lot of things extremely well. Unfortunately, few of them are valued as priorities by sports car owners. Courtesy Toyota Motor Sales.

Spyder to what *Sport Compact Car* called "spunky" performance, the 1ZZ-FE wasn't meant to break dynos and turn the world on its ear; it was intended to be lightweight, quiet, compact and fuel-efficient, produce low emissions and be inexpensive to assemble.

While none of the above sounds very ambitious, remember that the MR2 Turbo and third-generation RX-7 twin-turbo stalled on their own ambition only a generation earlier, producing exhilarating performance at a significant cost which ultimately led to low sales.

Toyota was eager to prove that it had learned its lesson from the MR2 Turbo—that performance doesn't automatically equate with product success—and that they could develop a sports car for the everyman, in sturdy, reliable, affordable Toyota fashion.

Just don't expect for it to stir your soul.

While Toyota continuously made revisions to increase chassis stiffness, the most significant changes to the MR2 Spyder involve the transmission, with a sequential transmission was offered in the 2002 MR2 Spyder. This was a significant achievement for Toyota, with the MR2

TOYOTA MR2 PERFORMANCE

The MR2 Spyder represented a triumphant return for the MR2 nameplate to the U.S. market. The early models can be identified by a front mouth that turns downward. Photo courtesy Toyota Motor Sales.

ZZW30 ENGINE SPECIFICATIONS
Block/Head Type: Iron/Aluminum alloy
Displacement: 1.8 liters (1794cc)
Bore & Stroke: 79mm x 91.5mm
Compression Ratio: 10:1
Cam and Valve: DOHC, 16 valves
Maximum Horsepower: 138 hp @ 6400 rpm
Maximum Torque: 125 lbs-ft. @ 4400 rpm
Redline: 6800 rpm
Induction: Normally aspirated
Recommended Fuel: Unleaded Standard
Fuel capacity: 12.7 gal.
Oil capacity : 3.7 qt.
Horsepower per liter: 76.67 hp per liter

Spyder by far the most inexpensive car to receive such an ambitious method of gear-change, sequential units to that point generally reserved for Ferraris and other cool cars the average enthusiast can't afford. It turns out, though, that this was an answer to a question nobody was asking, and it never jump-started MR2 Spyder sales the way Toyota hoped it might. And while an interestingly engaging driving gimmick, it also made the ZZW30 slower, noticeably reducing acceleration, with *Sport Compact Car* reporting quarter-mile E.T.s .8 seconds and 4.3 mph slower, a significant amount, and an edge the already tepid 1ZZ-FE couldn't afford.

In 2004, the MR2 Spyder was offered for the first time with an optional limited slip differential but by that point, the writing was on the wall. For the MR2, that meant five years, so by 2004 it was on borrowed time, and after 2005 it quietly bowed out of production.

THREE MR2S, EIGHT ENGINES

The AW11 will never be mistaken as a powerhouse, even in supercharged form. 1.6 liters just isn't very much, and short of a Formula-1 type effort, it's pretty much a lost cause as a performance engine. It is difficult to imagine, then, the AW15 making even less power than the stock 4A-GE offered—significantly less.

But it did. Toyota initially offered home market buyers a 1500S MR2, powered by the 1.5-liter, carbureted, 1452cc, 12-valve SOHC 3A-LU engine cribbed from the Toyota Corsa. In fact, throughout its colorful history, each engine the next generation MR2 received was significantly different from the last. The MR2 couldn't seem to make up its mind what it wanted to be, either true all-out sports car or everyman daily driver. Every few years, the MR2 either got a new engine, or a complete generation overhaul. There is something to be said for simplicity and linear change.

Across its three generations, the MR2 was equipped with eight unique engines, not counting the two rather significant revisions to the 3S-GTE. That is a large number for a car with only sixteen years of production in the United States, and only a handful more worldwide. The engines by code were:
3A-LU: (AW10)
4A-GE (AW15)
4A-GZE (AW16)
3S-GE (SW20)
3S-FE (SW21)
5S-FE (SW21)
3S-GTE (SW20)
1ZZ-FE (ZZW30)

Chapter 2
Building an OEM Healthy MR2

MR2s are unique in their ability to perform brilliantly as both daily drivers and on the track. While exceptional design and manufacturing by Toyota didn't hurt, the secret to keeping your MR2 healthy no matter where you use it is responsible, attentive ownership. Photo courtesy Tommy Guttman.

There are a variety of excellent books available that cover basic and advanced maintenance procedures, including the Toyota factory service manual or BGB (see sidebar on page 22) as referred to by MR2 enthusiasts. Rather than rehash standard maintenance procedures that are already well-covered in these books, I've chosen a handful of popular or performance-oriented procedures that supplement and improve on those guides where MR2 owners need them most. In this chapter specifically, our focus is on outlining what constitutes a "healthy" MR2, in the interest of long-term ownership, as well as establishing a strong foundation for future performance.

We'll look at several maintenance tasks, including compression testing, the basics of check-engine lights, and general component testing, to collectively establish a means to measure and maintain your MR2's health. We'll also address some of the body and mechanical maintenance issues more common to MR2s, including ignition and timing advice, gear-change troubleshooting, rust, gas mileage (directly affected by both maintenance and modifications), and the one of the most misunderstood items on the AW11 and SW2x, the airflow meter (AFM).

GENERAL MR2 HEALTH

Before you start modifying any car past its stock condition, it needs to be in good operating condition. This is true even if you don't plan on more than just a few mild upgrades. "OEM healthy" is a term I use to describe a vehicle that is in "mechanical condition at or approaching the level of tune as delivered from the factory, as close to stock specifications as possible." This applies to the engine, transmission, clutch, drivetrain, suspension, brakes, tires, electrical components and other vehicle systems. This is a list that can get long fast.

General 5-Point Check

1. The primary system for any engine tune is generally the ignition, usually spark plugs and ignition wires, and on many older engines, the distributor cap and rotor as well, including the base timing. These and the fuel filter should all be regularly replaced with Genuine Toyota parts, and timing should be set to Toyota specifications for your specific engine.

2. Tires should all match in brand and condition, something exceptionally important on the mid-engine, rear-heavy MR2. Also, while very obvious, the tires should not have any steel-belt (strands of metal) exposed. Legally, tires need to be replaced when the tread depth reaches 2/32". Insert a penny upside down into the tread depth; if the tread doesn't reach Abe's head, the tire needs to be replaced.

3. The clutch should be able to endure moderately aggressive launches without filling the cabin with the stench of clutch (an unmistakable, strong smell). It should also be able to provide low-rpm, high-gear\wide-open throttle runs.

Automatic-transmission equipped MR2s should be cautious for action similar to a slipping clutch. If the car lurches in gear, refuses to upshift or downshift at times, or slips under aggressive acceleration, have it inspected before you start your modification process.

Toyota MR2 Performance

The fuel filter is out of sight, but don't let it be out of mind. They are relatively accessible, and removal and replacement takes about 20 minutes. Along with ignition tunes and oil changes, these are the basics of MR2 engine maintenance.

4. The braking system should not make any loud grinding noises. This sounds like a crude assessment, but along with pedal feel and actual stopping braking performance, this is the easiest way to measure the health of your brakes. While some high-performance brake pads can get squeaky, grinding, screaming, clanking or other unnatural noises should be addressed right away—on any car, much less one you are modifying to go faster.

5. Suspension components should not be excessively worn. Ball joints are high wear items on MR2s. Suspension bushings and steering system components should all be in proper working condition as well, as should shocks. The alignment should be set to OEM specs. (More information on alignment is available in Chapter 3, and in the staged buildups sections of this book.)

Base Timing the 3S-GTE

Increasing base timing is an old-school way of finding some easy horsepower. This usually entails loosening and rotating the distributor a few degrees. With sufficient octane, advancing the base timing on a normally aspirated engine will usually make a few horsepower, in the neighborhood of 3–5 whp on a stock 4A-GE for example. And though this has been done for years, you have to be willing to risk detonation for a few horsepower. Even on normally aspirated engines, which typically are less sensitive to tuning than forced-induction engines, this isn't the safest, smartest or most effective way of increasing horsepower. On a turbocharged car, especially a 3S-GTE, it is a definite no-no.

The 3S-GTE ECU does funny things with timing to begin with, with OEM timing maps curiously off-the-scale advanced at points on the factory maps. Base timing, that is the timing set by the position of your distributor or other relevant

If your MR2 has cross-drilled rotors, always inspect them closely for cracks near the holes themselves, especially before any track or autocross event. Brakes are a safety item first, performance second, and should take the highest priority if you are suspicious there might be issues.

3S-GTEs are particularly hard on ignition components, and sensitive when they are out of specification.

component, seems accessible and tempting enough to fiddle with, but on the Gen II 3S-GTE in your SW20, set it between 8 and 10 degrees, and leave it alone.

The Compression Test

A compression and a leakdown test can give you even more information about the state of your MR2's engine. Before starting any bottom end modifications, and even intermittently thereafter, a compression and leakdown test is a good idea.

Most cars maintain good even compression for years, but if you're running higher-than-stock boost levels, or actively racing your MR2, it should be done with greater frequency.

Well-designed production cars possess an impressive amount of flexibility in their power output. With appropriate engine management—and base timing—the 3S-GTE's stock rods, crank and pistons have the strength to withstand over 400 hp consistently.

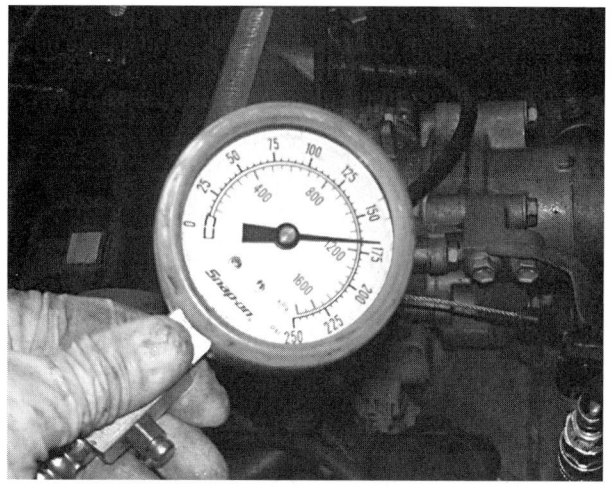

A compression test is a good starting point in measuring the health of your engine. The more data, the better. Photo courtesy Jensen Lum.

A compression test is basically a good guess of how healthy each combustion chamber is—how well it seals and squeezes air. The better it seals, the better the health of the sealing components, namely the piston rings and valves. The better the health here, the more efficient your MR2's engine will operate and the more power it will produce. Basically, low compression readings mean the piston isn't squeezing and holding the air very well, appropriate compression will usually mean everything is working as it should, while high compression is generally indicative of either carbon deposits or aftermarket pistons. To see what kind of compression you've got going on, the following test applies:

Compression Test Procedure—On a warm engine, perform the following steps:
1. Remove all four spark plugs.
2. Open throttle-plate in the throttle body, either with the gas pedal or by hand
3. Disable the ignition system, by pulling the igniter/coil wire for example.
4. Disable fuel system, by pulling the EFI fuse, or each individual injector connector, for example.
5. Screw in the compression gauge into the #1 cylinder's spark plug hole, and crank the engine about 3 to 5 times.
6. Repeat the test for each cylinder, writing down the results for each cylinder. It's easy to forget.

Analyzing Compression Numbers—So you buy or rent a compression test kit at your local automotive store. You perform all of the steps, crank the engine and watch the gauge—what kind of numbers are you looking for? It depends.

The BGB for your MR2 will give you the exact numbers you're looking for, because as compression varies across variations of the 4A-GE, 4A-GZE, 5S-FE, 3S-GTE, and 1ZZ-FE, so do appropriate compression numbers. When performing a compression test the most important thing you want to see is consistency. While low compression numbers are certainly cause for concern, seemingly high or low readings may have explanations that do not involve worn piston rings and blowby. For example, aftermarket pistons or headwork can play a factor in your engine's compression, and if you purchased your MR2 used, as most owners do, this could be worth checking out.

Ideally readings should not vary more than 10% between cylinders, so an engine seeing 150 in one cylinder should not see less than 135 or more than 165 in any other. If you are indeed seeing one cylinder more than 10% off from the rest, or simply have low compression everywhere, a second opinion—as in another confirming test—is in order. If the worrisome numbers are confirmed, your engine needs to be checked out by a professional, and probably rebuilt.

Compression Numbers and Carbon Deposits—A common scenario on the SW20 has the owner testing and finding very high numbers, as high as 210 across all four cylinders. Usually this points to carbon deposits. Carbon deposits are basically combustion residue from the convergence of oil, fuel, and extremely high temperatures in the combustion chamber.

And this is not a good thing. Not only do you want your compression ratio static and unchanging, but you want the temperatures across your pistons as even as possible. If high compression numbers

POTENTIAL CAUSES OF LOW OR HIGH COMPRESSION READINGS
- Faulty test procedure or equipment
- Carbon deposits in combustion chamber
- Blown head gasket
- Worn piston rings
- Physical internal damage to piston or combustion chamber
- Stuck or otherwise damaged valve
- Timing belt off-time

TOYOTA FACTORY MANUAL: "BGB"

Toyota has a guide that has many of the maintenance procedures your MR2 will require listed in step-by-step format, including the above compression test. This guide is called the Toyota Factory Service manual, affectionately referred to by many MR2 owners as the "BGB" an abbreviation for big green book—because it's big and it's green—and is something you'll need to familiarize yourself with. An MR2 owner's bible, this manual is priceless, though officially you can expect to pay anywhere from $50 used, to closer to $225 for both volumes new from many Toyota dealers. It contains every how-to, torque rating, capacity and overall specification you might ever need, though it is not without the occasional questionable reference. The BGB can be ordered on the Internet directly from Toyota at www.techinfo.toyota.com/public, or by calling 800-622-2033. If neither of those sources is available, an Internet search or your local Toyota dealer is your best bet. Alternatively, you can download pdf files on a subscription basis by visiting www.techinfo.toyota.com.

---

are what you're seeing, consider water injection, which keeps these deposits from forming, and will rid your engine of those that have already formed.

Wet Compression Test—If the results are low, pour no more than one teaspoon of oil into the lowest-reading cylinder, perform another test and see if that raises the compression. If it does, it most likely means your piston rings are excessively worn, and the bottom end of the engine should be rebuilt or replaced. If it doesn't affect the compression reading at all, most likely the valve is sticking open, or isn't seating all the way, or the head gasket is damaged.

To reiterate, you are not looking for high numbers, but consistency, 146-145-144-147 being more desirable than 170-175-85-175. But that's not the whole story.

Consistency is indeed good, but may simply be evidence that something amiss is at work across all of the cylinders evenly. While consistent numbers do makes it less likely that there is significant damage within the combustion chamber of any one cylinder, it doesn't speak for your head gasket, for example. You have pistons attempting to create compression, but that compression is escaping somewhere. To find out where, you'll need another test.

The Leakdown Test—In some cases it is possible that the compression test reflects perfectly normal numbers while still concealing damage to the engine somewhere. A 3S-GTE could show 155 across the board, apparently indicating a healthy engine, but that 155 could be artificially produced by something other than mere piston-to-cylinder head squeezing. To add to your compression data, you also need to test the engine's ability to hold injected air.

A leakdown test allows you to more closely pinpoint where an issue might be, rather than measuring compression, instead beginning to identify *where* the compression is escaping. A compressed air tank is necessary, as is a leakdown test kit, which is more expensive than compression test kits. The engine must also be placed at TDC (top-dead center), all converging to make the leakdown test more complicated than a compression test, and easier to garner inaccurate data. Any good leakdown test kit will come with instructions, and that valued mechanic you've hopefully established a working relationship with can almost certainly perform one as well.

Check-Engine Lights: AW11 AND SW2x

The ECU in your MR2 is smart, and it has an excellent memory. Not only is it responsible for orchestrating the concert that is a running combustion engine, but it also can tell when something is wrong. Sometimes it gets things wrong, of course—hears a knock when there may be none, or tells you your O_2 sensor is shot—again—when it's really fine. But it's all we've got, and overall it's very reliable.

The ECU is connected to a multitude of sensors scattered throughout the car, including a whole slew in the engine-bay, reporting back to it throttle position, oil pressure, coolant temperature, rpm, audible detonation events, and a number of other

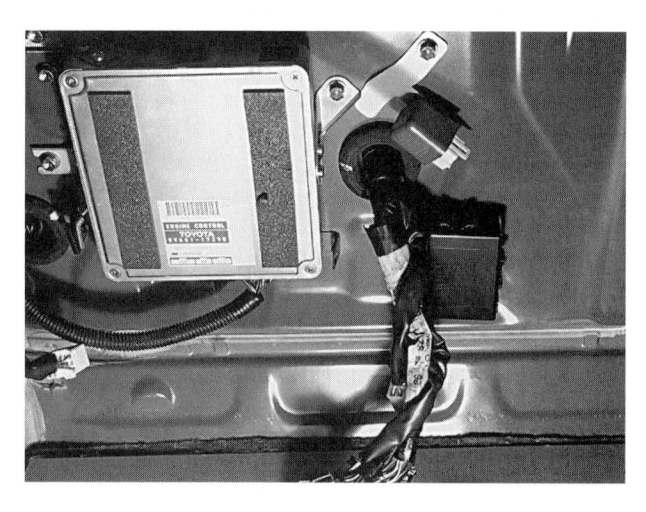

The ECU is the car's central command system. Dozens of sensors constantly report to the ECU through wires that are bundled in various looms. These looms finally converge at the ECU in the rear trunk.

conditions that help it decide how to most effectively accelerate the vehicle while keeping the engine from damaging itself. When one of the sensors is claiming something is wrong—or the sensor itself is failing—the ECU will, depending on the severity of the event—trigger a check-engine light, which illuminates on your instrument panel in some distressing hue, usually red or gold.

Note: The ECU can also store codes without even showing a check-engine light, so checking and clearing codes regularly—with oil changes on a car seeing intense application, annually on MR2s seeing less extreme use.

Checking for Codes—To check for stored codes, grab a BGB and following the instructions located within is your best bet, as always. As a rough guide, though, consider the following procedure for the SWx2:

1. Be sure the battery is fully charged, and that the throttle is fully closed (no pedal pressure). Turn off the A/C, and warm the engine to normal operating temperature.
2. Turn the ignition switch to ON (without starting the engine)
3. Using SST 09843-18020, or an uncoated metal paper clip, whichever you can find first, connect terminals TE1 and E1 of the diagnostic connector.
4. Watch the check-engine light on the instrument panel, and count the number of blinks. If it simply blinks on and off consistently, there are no stored codes. Otherwise, the light will blink twice per second. Counting the number of times it blinks will give you the first digit of the 2-digit code. Following a 1.5 second pause, the second digit will be similarly revealed.
5. If there is more than one code, the next code will be revealed after a 2.5 second pause.
6. The codes will continue to blink as long as the terminals are connected.

Note: Keep in mind that the just because the ECU flashes the code doesn't necessarily mean that the corresponding part listed below is faulty—a code is generally just the first step in troubleshooting an issue.

Clearing Codes—If you get a check-engine light and pull a code 32, for example, and you aren't convinced something is in fact wrong with the AFM, you can always clear the code by pulling the fuse to the ECU or disconnecting the battery for 60 seconds. This will clear all stored codes, so if it comes up again, you can now be certain that the ECU at least thinks something is wrong with the AFM. Keep in mind, though, that disconnecting the battery will also cause you to lose your clock and radio settings.

Testing Engine Components

Component testing doesn't sound very exciting, but it can save you a significant amount of time and money in troubleshooting. There are parts on all cars, including MR2s, that fail with some degree of frequency, and knowing how to test those parts can save you a great deal of time and money, not to mention stress. The AFM is one of those parts.

Typically, if you're having problems, replacing a failed part with a new or reconditioned one is the best course of action. Sometimes this is not possible, however. In cases like these, if possible actually testing the component itself can be helpful.

One way to test components is with an ohmmeter. An ohm is a unit of resistance—the higher the number of ohms, the more resistant a circuit (of any sort) is to allowing current flow. If there is no resistance in a circuit, the ohmmeter will read 0; an open will read "infinite," or whatever matter of reporting such a status your particular ohmmeter uses.

There are digital multimeters that incorporate several such functions of current and resistance metering in a single unit, usually about the size of a deck of cards and battery powered. Digital multimeters can get very complicated, and very expensive, able to measure everything from injector pulse-width to temperature and every form of energy. For now, we'll stick to the old-fashioned analog ohmmeter.

In testing a circuit, there should be no power in that circuit to minimize the potential of damaging

TOYOTA MR2 PERFORMANCE

ZZW30 DIAGNOSTIC CODES

The codes from the 1ZZ-FE in the ZZW30 are considerably more complicated than in the AW11 or SW2x. The given codes are far more specific than in previous generation MR2s, able not just to diagnose a misfire, but which cylinder misfired; not just failures, but degrees of failure.

For example, sample of 1ZZ-FE diagnostic codes:
P0201 Injector Circuit Malfunction—Cylinder #1
P0202 Injector Circuit Malfunction—Cylinder #2
P0203 Injector Circuit Malfunction—Cylinder #3
P0204 Injector Circuit Malfunction—Cylinder #4

As you can see, though horsepower production has decreased, engine development has gotten more precise over the years. Overall, there are hundreds of available codes on an MR2 Spyder, so a BGB or a certified mechanic is ideal to troubleshoot ZZW30 engine codes.

Note: If you a previous owner has "swapped" or otherwise done significant repairs on your MR2, be sure to check whether or not your check-engine light, and other warning lights, are even in fact "plugged in" and operational. Many owners, or contracted speed shops unbeknownst to those owners, take shortcuts in doing swaps, and simply render "annoying" check-engine lights inoperable.

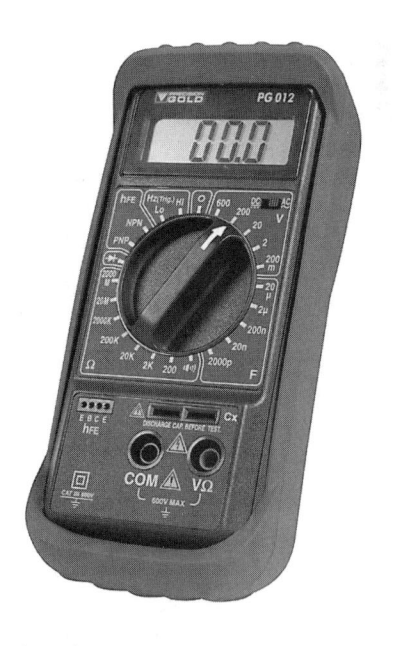

An ohmmeter, or a meter that measures electrical resistance, is available with a multimeter, a larger device capable of a variety of measurements, usually made through leads (not shown). This multimeter is the GW20, manufactured by Precision Gold. Photo courtesy Precision Gold.

the ohmmeter, so if you're using a multimeter, make sure it is set on what you are trying to measure. The ohmmeter allows for specific parts and inherent circuits to be tested to gauge their relative health—sort of like a doctor taking a patient's blood pressure. Sometimes parts fail slowly, and may "look" fine, that is they aren't on fire, split in half, or black with electrical burn.

Note: There is some disagreement on whether using an ohmmeter to test O_2 sensors is a good idea or not, the concern being that the testing procedure itself might damage the sensitive circuitry of the O_2 sensor. I have never had any issues testing them, but there are some that swear that such damage is possible. Be aware.

The Airflow Meter—The airflow meter, or AFM, fails occasionally. Sometimes it kinda sorta works, sending out a signal, but a faulty one. This can cause the ECU to use erroneous timing information, possibly damaging your engine. The AFM is right in the line of fire between the atmosphere and the turbocharger, and gets dirty fast. It also is a spring-loaded part more than a decade old and counting, so issues should not be completely unexpected.

Oftentimes though, these issues are not because it is poorly designed or manufactured, but simply because too many people tamper with it. While a few 4A-GE owners in recent years reported success by cutting off the top of the AFM and adjusting the spring tension to lean or richen the mixture, this is a practice that I do not endorse. If you know what you're doing, and you've done it before, maybe. But for the majority of owners out there looking for power, increased fuel efficiency or performance, any adjustment of the AFM is not worth the potential headache.

So what happens when someone has tried to adjust the AFM and damaged it? Or perhaps your engine is not running quite right, and everything else checks out fine? It is time to break out the ohmmeter and give the AFM a once-over.

Removing and Testing the AFM—Before you can test the AFM you need to unplug it, and that may appear more complicated than it really is. The AFM has screws on the side that, for some reason, people can't help themselves from removing. Anyone that has pulled them knows the horror that ensues; whatever you do, do not remove the screws from the AFM. The AFM can be disconnected by simply removing the spring clamp on each side, and unplugging the connector.

Note: Depending on your setup, removal of the entire AFM may not be necessary to test it. If you want to clean it (with light-duty cleaner, and/or a dry rag), or replace it with a working reconditioned or new unit, then obviously you'll have to remove it.

Note: The procedure in the sidebar on page 26 purposely visually demonstrates the 3S-GTE AFM removal, and highlights a 4A-GE AFM testing. For most intents and purposes, an AFM is an AFM, and testing one is very similar to testing another. Since most owners are 4A-GE or 3S-GTE owners, these quick steps give attention to both setups.

24

BUILDING AN OEM HEALTHY MR2

The small-port 4A-GE, distinguishable via the ribs on the top of the intake manifold plenum among other methods, was also available in MAP form, removing the AFM entirely. If you look all the way to the right, you can just see that this small-port 4A-GE, out of temporary necessity, was actually using an AFM, along with the large-port ECU and tangent electronics.

4A-GE TERMINALS

From left to right, the AFM terminals on the 4A-GE are recognized as:

E1 FC E2 VB VC VS THA

In testing the resistance between the terminals, the results should fall somewhere between the given specifications:

Across Terminals	Specified Resistance (ohms)	Temperature (degrees F)
E2—VS	20–3000	N/A
E2—VC	100–300	N/A
E2—VB	200–400	N/A
E2—THA	10–20 k	-4
	4–7 k	32
	2–3 k	68
	0.9–1.3 k	104
	0.4–0.7 k	140
E1—FC	Infinity	N/A

The airflow meter, or AFM, has been known to give MR2 owners fits. If out of specification, it can lead to running problems exceptionally hard to diagnose. A multimeter can save you some time here. (3S-GTE AFM shown.)

Use the guide, and what you see in your engine bay, cooperatively with the BGB to figure the simplest, safest way to perform the procedure.

For AFM electrical reference, we will consider the AW15's 4A-GE AFM. Looking at the 4A-GE AFM connector from the side, there are 7 prongs, each a terminal by electrical standards. To see if our AFM is functioning properly, we can test the resistance between each terminal with our ohm setting on a multimeter.

Each terminal has a corresponding alphanumeric code so that it can be read in wiring diagrams and schematics—for times like now (works much better than "third one from the left").

Vacuum Leaks

I once drove over a thousand miles to buy a gorgeous 1991 Steel Mist Grey SW20. For the first 500 miles of ownership, everything was fine. Unfortunately the car I was purchasing had the GReddy TD06SH, which suffers from horrible boost creep. Boost creep is basically when a turbocharger continues to build boost beyond what it is set to produce. Usually this occurs due to the wastegate's inability to bleed off sufficient air to keep boost at safe levels. Boost would sometimes spike to 23 psi or higher, which felt pretty cool to be honest, but was definitely a not-cool thing.

I had a local shop install an external wastegate to control boost, and had them install a tubular exhaust manifold at the same time (welding external wastegates to cast iron stock manifolds doesn't always work out well). Funny thing, though—after the install of the wastegate and manifold were done, the car wouldn't run right. Rough idle, wouldn't boost properly, etc. What had I touched last? The turbo kit. Best to start there.

It certainly seemed like a textbook vacuum leak, only I couldn't find it. Neither could said shop that installed the wastegate/manifold assembly. I performed the typical tricks to find a vacuum leak, namely spraying carb cleaner (carefully, away from the hot stuff) in hopes of temporarily raising the idle and locating the leak. That yielded nothing, as

REMOVING THE AIRFLOW METER

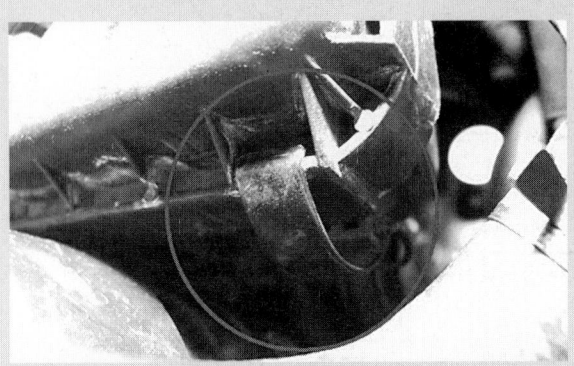

1. On the 3S-GTE, to remove the upper portion of the filter housing, the fasteners that marry the upper and portions of the filter housing must be undone. (3S-GTE shown.)

2. The clips are easy to remove—a tiny flathead, or other non-bludgeoning device that can be used to pry the clip just so, will work here.

3. After the airbox fasteners have been undone, move on to unclipping the AFM plug itself. Do not remove the screws. (3S-GTE shown.)

4. Once unplugged and disconnected from the lower half of the air filter housing, the AFM will look like this. Notice the screws intact and undisturbed. (3S-GTE shown.)

did other garage tricks that can sometimes track them down: spraying water on hoses, listening as carefully as possible for leaking air, etc. For four months I literally could not drive my new car.

I went back over everything I had changed in the conversion to an external wastegate and tubular exhaust manifold. Was it really a vacuum leak? Surely we had forgotten something—missed something obvious. Maybe a sensor coincidentally had gone bad—but everything checked out fine: resistance was perfect on all pertinent sensors, and the AFM checked out fine as well. Compression and leakdown tests came back perfect, so the bottom end of the engine wasn't the problem either. I checked the timing five or six times, and even suspected the timing belt may have jumped off a tooth (though when that happens, it will usually render something worse than what I had going on),

but that was all perfect as well. I needed to verify that it was indeed a vacuum leak, so Bryan Moore suggested I perform an "AFM vacuum-leak test."

Vacuum Leak Verification and the AFM—Considering that the air is measured several feet before it enters the head, there is ample opportunity for poorly sealing air-moving and transfer mechanisms (for lack of a better term) to leak air. If the AFM is telling the ECU that X amount of air is being ingested, and only X minus some given amount actually makes it, you'd think the car would run rich. Well, that'd be right if it weren't for the vacuum condition that exists until boost is achieved. Remember, vacuum "leaking" is an inaccurate term: "sucking" is more appropriate.

Instead of rich, the vacuum leak will allow even more air to enter the cylinder, and the car will run lean. The O_2 sensor will attempt to adjust for this

The offending SW20 after I had sold the $2,000 Volk Racing wheels to help pay for the exhaust manifold, coating and external wastegate. Body kits and stock offset/wheels don't mix.

In addition to a magically disintegrating downpipe and poorly sized wastegate that allows dangerous boost creep, the GReddy TD06SH also suffers from an adapter that is not properly port-matched, inhibiting airflow. If you've got a TD06SH kit, port-match these pieces before you install.

A vacuum leak can be difficult to find. Can you spot the vacuum leak? Neither could I, after no less than six hours using standard vacuum-leak-detecting methods. (Hint: It's not the missing intercooler pipe).

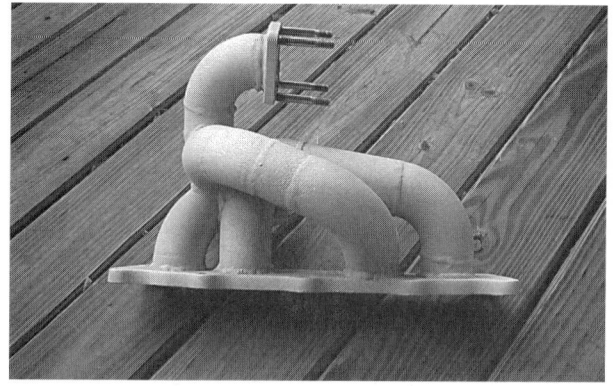

A tubular exhaust manifold is one way to enable the use of an external wastegate to bleed off boost (as opposed to the internal wastegate of the TD06SH, and most stock turbochargers). Care must be taken where the external wastegate is plumbed, however, as improper placement in an area that sees too little exhaust flow could keep the wastegate from functioning correctly.

condition at idle, partially disguising the problem. Under boost, though, the car will now run rich, with the air-fuel mixture being thrown off in the other direction as the car refers to the ECU fuel and timing maps under WOT (wide-open throttle).

Back to the vacuum leak: if the engine is running poorly, especially when cold, there is a trick you can use to help determine if your problem is a vacuum leak, or something more expensive.

1. Remove enough of the air filter assembly to gain access to the flapper door on the AFM (depending on your engine, and how your MR2 is set up).

2. With the engine running, *barely* push open the door flapper on the AFM, being very careful to make sure no debris enters the AFM. This will suggest to the ECU that more air is being ingested, which will convince the ECU to add a little fuel.

3. Check the idle. If you have a vacuum leak, it should smooth out considerably.

4. Double check by allowing the flapper to fully close—if the smoothed-idle becomes rough again, you very well may have a vacuum leak somewhere.

Note: Those of you with Gen III 3S-GTEs, or full JDM setup small-port 4A-GEs, you're running a MAP sensor to meter air, and vacuum leaks aren't such a soul-crushing thing.

After performing the above test, I knew I had a vacuum leak. Now at my wit's end, I flew a very skilled Toyota Master Tech by the name of John Reed—ironically a good friend of the gentleman who sold me the car to begin with, and the man who had built the engine—across the country to find the problem. In about five minutes, with the

Convincing the ECU to add fuel is as simple as just slightly pushing the flapper door open on the AFM. If the idle smoothes out, chances are good you've got a vacuum leak.

Worn shifter cable bushings can lose precision, sometimes keeping your shift lever from properly engaging gears. This is especially true of older, higher mileage bushings, or MR2s located in high-corrosion areas. Start with an inspection of your current bushings. A new set of bushings just might do the trick.

same can of carb cleaner I had used, he found a missing bolt from backside of the 20G turbocharger housing that was allowing a monster vacuum leak.

A missing bolt. That's all it took. Through process of elimination, use of the AFM as a diagnostic tool, and the humbling realization that there are people out there who are smarter than I am, my vacuum leak was repaired, and I was able to drive my MR2 once again.

Transmission Health

Another common issue facing MR2 health is shifting performance. MR2s, all aging and cable-shifted, are often critcized for their shifting performance. But poor shifting performance does not necessarily mean your transmission is on the way out. It can mean that your gear oil needs filling or replacing. If you've changed fluids recently and still have problems, find another fluid. Gear oil is not the place to save money.

It's also possible your MR2 has simply had a poor-quality aftermarket short-shift kit installed at some point in its life. Oftentimes such kits can place odd stresses on the gearbox, so verify you're working with a stock shifter, or that your aftermarket kit isn't the source of your problem. If neither of these pan out, the remaining potential issues can get expensive fast. This list starts with a worn clutch.

Road-Testing Your Clutch—Many aftermarket clutch setups are not the most peaceful bits of machinery in the world to operate, nor are they as reliable as an OEM clutch. And when they act up or fail outright, the problem generally cannot be diagnosed or repaired without pulling the entire clutch assembly—which means the transmission, too. Have fun with that. A failing clutch can also act a lot like a worn transmission. With a fading clutch, the shift lever will resist going into gears, rapid gear changes can be met with grinds—overall the transmission will seem disinterested in doing what it's supposed to do. So how do you know what's causing your notchy shifter: low or old gear oil, worn synchronizers, a failing clutch, or just your imagination? One way to narrow it down once you've verified you're working with fresh gear oil is to test the clutch's torque holding.

Torque is what clutches are paid to hold, not horsepower. While it would seem that launching the car hard from a stop would put the greatest strain on the clutch, it's actually the high-load low rpm-high/high-throttle input that really stresses it the most.

The easiest way to test your clutch is to find an open stretch of road, accelerate to around 2500 rpm in a higher gear, say 4th. Peak torque on a stock MR2 Turbo is only a few hundred rpm beyond that, and that's what we want—peak torque. In a safe location, at 2500 or so rpm, apply wide-open throttle. If the clutch is healthy, it will hold, and the rpm and speedometer will both accelerate in a similar fashion. If the clutch is failing, the engine note will rise artificially fast, but the car will only slightly accelerate, if it does at all.

If it works the first time, try it two or three times to be sure. Clutches are shady characters, and don't want to be replaced; you've got to watch them.

Adjusting Your Clutch Pedal Freeplay—One potential to explore before tearing out your transmission is the adjustment of your clutch pedal freeplay. Nearly all Toyota models are adjustable for clutch pedal freeplay, including the MR2. An out-of-adjustment pedal can keep the clutch from fully engaging and disengaging, which puts a lot of wear on the clutch disc, and, consequently, on your transmission synchronizers as well as you row, with increased force, through the now-stubborn gear sequence.

Also, many times a new clutch, especially shoe aftermarket jobs, even with the hydraulic system properly bled will not yield a "good pedal feel," and pedal freeplay must be adjusted periodically as the clutch disc wears. I have even seen very stiff aftermarket clutches—as they typically require extreme amounts of pedal effort—eventually wear out the clutch pedal, literally forming a groove in the metal on the backside of the pedal "arm" near the firewall, removing movement from the potential arc of engagement, and not allowing the clutch to fully engage or disengage. (This is also a good time to remind you that your clutch pedal is not a foot rest—that is what the dead pedal to the left of your clutch pedal is for. Resting your foot on the clutch pedal between shifts is an easy habit to form, like any self-respecting bad habit should be, but is equally harmful.)

The following steps are designed to supplement the BGB procedures.

1. Measure the pre-adjustment freeplay of the clutch pedal. This is done by using a ruler or measuring tape, and literally measuring the distance from where the pedal rests, to the point at which the pedal begins resisting further movement (it is trying to engage). Check the BGB for correct freeplay specifications for your specific MR2. The clutch pedal should also not move more than a couple of millimeters laterally, if at all. If the freeplay is not within the stated range of movement, it requires adjustment.

2. Way, way down on the floorboard, underneath the dash is the business end of the clutch pedal, where it is anchored to a pushrod (one of the few places in any proper sports car where a push rod belongs) that actuates the master cylinder on the other side of the firewall. There is a locknut on the end of this pushrod—that's where the adjustment occurs.

3. Loosen the locknut, and twist the push rod until you're within the given tolerances.

4. There is also an adjustment for pedal height just above our rod/locknut assembly, a little further up the pedal. Check and, as necessary, adjust the pedal height now.

The more freeplay you have, the lower your clutch will "bite," and the less freeplay, the higher it will bite. Too low an engagement can cause wear on the gears though, and too high can allow for too much clutch slip, prematurely wearing the clutch assembly. In the end, you should adjust it so that it

> ## SHIFTING PROBLEMS?
> If you're having trouble shifting your MR2, before you start budgeting the cost of a new transmission, perform the following steps:
> 1. Make sure your transmission fluid is topped off and fresh; if still an issue, try a performance fluid like Redline MT90.
> 2. Road test clutch to verify its condition. A failing clutch will keep the transmission from operating properly.
> 3. Adjust clutch-pedal freeplay. See above.
> 4. Verify shift lever operation. Short-shift kits, worn shifter bushings, and other low-dollar items can combine to keep the transmission from engaging properly.

engages to suit your driving style.

Note: Our cramped little working area here is also where the clutch start switch is activated. The clutch start switch ensures that the clutch pedal is depressed before the engine is started, a safety feature that, if your MR2 is used, might have been disconnected by the previous owner. Check and verify that your clutch start switch is intact, and works. Beyond eyeballing that it is indeed there (it's a cylinder-shaped object near the top of the clutch pedal) the easiest way to test is to try to start the car without pushing in the clutch. If it starts, you now know where to look.

MR2 EXTERIOR

Your MR2's exterior is usually thought of merely in terms of aesthetics, but, like your own skin, it is also entirely functional.

Not only does the exterior protect you in the event of a collision, but it also provides a structure that dictates structural rigidity, and center of gravity as well. This in addition to providing the hopefully flattering aerodynamic characteristics for gas mileage as well as performance Toyota gave it from the factory. When factors, such as accident damage, corrosion and rust compromise these characteristics, performance—and more importantly, long-term ownership—will certainly be sacrificed.

Rust

A discussion on keeping your MR2 healthy would not be complete without some attention given to rust. Rusting has to be among the most unfortunate of chemical processes, not quite the harbinger of dread that is nuclear fission, but more commonly irritating.

It really is a very awkward process: one day there

Toyota MR2 Performance

AW11s tend to rust worst on the rear quarter-panels, from the inside out. Not a sandpaper-and-paint proposition, this 1989 MR2 is going to need new fenders.

Remove your fender liner and have a look at what might be clogging those rain drains. Beware wombats. Photo courtesy Aaron Silvestri.

is painted, polished metal, there for you to touch, and then two years later only a space for you to put your hand through, the months in between marked by increasingly brazen bubbles of dying, painted metal rising up in some slow, dramatic death. The car simply begins to disappear in patches—over the rear wheel wells, sometimes just above the outside trim encasing the windshield, under the front lip on AW11s, or hiding elsewhere within like the coward that it is. Rust is wicked bad, depressing stuff.

It also seems to be somewhat inevitable. Some MR2s do seem more predisposed to rust. Many of these MR2s live in colder climates—cold begets snow, snow begets salt, salt begets corrosion and rust. The corrosive salt from the ocean is no friend to the MR2, either. Thankfully, though, the overpopulated west has huge areas relatively unaffected by either, where MR2s simply pile up mileage in heat and traffic, but tend to not rust.

To be certain, rust is the enemy to any long-term car ownership, but especially so in a car you dearly love, and want to own for the long haul. Though I love cars in a possibly unhealthy way, I am not one to jack them up, remove the wheels, fill the gas tank with rosewater and seal the entire spectacle up in some iron lung for the winter. While I do see the utility of a beater, unless it is a 2000GT, or there is snow actively falling on top of a salted road, my own rules of ownership state that I am driving my MR2 year-round.

To help resist rust though, there are some steps we can take.

Keep Your MR2's Exterior Clean—This is the single most important thing you can do it insure the long-term health and beauty of your MR2's body, and to a surprisingly large degree, its mechanical health and performance as well.

Keeping the body free from grit, dirt, and grime is essential for not only maintaining its sheen, but its value and, ultimately, its structural integrity as well. Not only should you wash your MR2 regularly—how often depends on your climate and driving conditions—but you need to be waxing and polishing it, too, and keeping the underneath clean and free from caked dirt, salt, and other stuff Toyota didn't put there.

Also be sure your underbody panels are intact as well, so that they may do their jobs. Clean isn't just for vanity's sake.

Keep the Rain Drains Free from Clogging Debris—Most cars, your MR2 included, have been provided paths over and through their sheet metal, and strategically placed drains to evacuate standing water. When these drains and pathways become clogged, water is allowed to pool, and rot your MR2 from the inside out. Rust is more common on AW11s, but is no less than threatening to your SW2x (though the removable body pieces sure make it easier to approach on the ZZW30).

If you can help it, be careful which kinds of trees you park beneath regularly. Not only do some trees bleed sap onto your paint and house birds that do worse, but pine needles and other smaller leaves are better at falling on your MR2 and finding their way into these holes and pathways. Shade is great, but keep an eye on that paint.

Don't Drive in the Snow—or store it for the winter. I'm assuming here that if it snows with any regularity in your city, they lay down salt, sand, or some similar grit (though in my neck of the woods the amount they lay in my town seems inversely proportional to the amount of snow that ends up falling). Either way, while driving in mere cold snow won't hasten the rusting of your MR2, driving through salt will.

As for storage, it ultimately improves the pool of existing MR2s when people do so. Storage can also introduce other threats to your MR2 if not done properly, so if it's something you want to do, educate yourself on the process well before proceeding. Simply storing your MR2 in a garage when it's not being driven can help preserve its condition tremendously as well. The sun is tremendously damaging to paint, trim and interior pieces.

Keep rust in perspective. On your MR2, rust is cause for concern and action. On a 2000GT, the most collectible Toyota ever produced, it is closer to Shakespearean tragedy. Photo courtesy Toyota Motor Sales.

Chapter 3
Basic MR2 Performance Modifications & Theory

With an open roof, a mid-engine, rear-drive design and a variety of potent powerplants, MR2s are blank canvases to create your vision of what a sports car should be. Just have a plan.

This chapter covers basic MR2 performance modifications and in it you'll find broad categories of automotive performance theory discussed, especially as they relate to the MR2 and its inherent strengths and weakness. This will lay the foundation for the more intermediate and advanced performance theory in Chapter 4, and the staged buildups of the platforms that follow in the proceeding chapters.

Once you have this background, you can really choose the level of performance you need. But in order to do that, you need to have a basic understanding of what parts do what, what adjustments and modifications are available, and how it all relates to your MR2.

The parts and modifications discussed in this chapter may seem very basic, but their impact on performance is substantial. These mechanical adjustments will give you the opportunity to experience your MR2 at another level, while continuing to learn more about automotive performance before you commit to significantly altering your MR2—and spending major dollars. Should you decide on advanced modifications to your MR2, you will build upon these basic, fundamental modifications.

BASIC HANDLING PERFORMANCE

Tires
Without really good gripping tires, you're not going anywhere fast. Launching, and aggressive braking and lateral maneuvers all depend on the available grip at the only four points the car actually contacts the driving surface. Increasing this grip returns exponential gains in performance, causing your MR2 to feel more capable the whole time it is in motion. Tires are that critical.

What Tire Is Best For Your MR2?—Bridgestone, Kumho, BFGoodrich, Toyo, Pirelli, Dunlop, Goodyear, Falken and others all make good tires; the best choice for your MR2, though, depends on a variety of factors. Your climate, driving style, the kind of racing you do—even the size of your wheels—all factor in considerably. The difference in dry grip between one tire and the next might be barely measurable, but the difference in wet grip might be downright considerable.

Tire Feedback and Feel—Sidewall height and stiffness is critical to overall feel and performance, with shorter and stiffer sidewalls offering sharper reaction and immediate feedback, but often at the cost of communication. Sidewall height and stiffness and the rubber compound of the tire both contribute significantly to how much warning the tires give before breaking away. This is a microcosm of the handling traits MR2 designers discussed in their development of the MR2; the higher the handling limits, the less warning the driver often has once those limits are exceeded. High-performance tires offer unbelievable amounts of grip, but—many of them anyway—are flat out gone once they "go" (that is, 100% of available grip is surpassed and they begin to slide).

Tires with stiff sidewalls will generally offer sharper turn-in and handling response because there is simply less tire to compress under the load of a turn. I personally love the response of a very stiff sidewall, as compared to the mush in

BASIC MR2 PERFORMANCE MODIFICATIONS & THEORY

In order for your MR2 to take its rightful place on a race track you must first understand which parts are responsible for what performance, and then begin to prioritize things based on your vision for your MR2. Photo courtesy Tommy Guttman.

The Toyo R888 is an excellent track tire. Note the symmetrical tread-pattern and the outer shoulder of the tire, which is rounded off to help maintain optimal tire contact under a variety of unique suspension loads. The R888 is an R-Compound, D.O.T. Approved Competition Radial. Photo courtesy Toyo Tires.

The Falken Azeni Sports shown here are fantastic, affordable tires. Falken and Kumho have helped pioneer low-cost, high-performance tires in sizes accessible for smaller cars like the MR2. Photo courtesy Falken Tires.

standard tall, soft sidewalls. When I am driving on a set of low-profile tires, though, I adjust my driving and expectations of the tires' reaction accordingly. There is a significant difference in feel, response, and communication between low and standard-profile tires. However, what tire is "best" depends on application. If I am simply zipping (safely) around town on on-ramps and city streets, a low-profile tire looks good and feels great. On a road course, low, stiff sidewalls work—and feel—even better, 35 or 40-series profiles for example. In these situations, removing all of the slack in the chain, so to speak, gives you the most responsive setup available, leaving everything up to the driver. This works well in these situations, when road undulations and direction change are entirely predictable, and pavement types, road debris, other factors that affect the friction between your tire and the road are known. On a mountain road, though, I'd prefer a taller tire that can more confidently soak up road irregularities, absorbing bumps without upsetting the suspension any more than it has to, and be a little less antsy when the grip gets sparse. In fact, in any environment other than the track, taller sidewalls, 50 series, for example, may be ideal.

This is similar to the stiff suspension versus softer suspension argument. Stiff suspension certainly controls roll and pitch better, but at what cost? On what surfaces do they perform better? Bowling-alley smooth roads, or the real roads you drive your MR2 on every day? The linear, gradual breaking away—and recoverability—of an understanding performance tire is not only preferably for the novice performance driver, but is not a bad choice for anyone. Find a tire that communicates to you what is going on in a language that you can understand. An MR2 that can be consistently and confidently driven at 98% will generally be faster than one ultimately possessing more grip, but driven with less confidence by the driver.

And don't forget that tire compound plays a significant role in it all as well, as does your exact suspension setup, and each must be considered carefully when selecting a tire, or a driving style that matches the tires you've got. Handling is a system.

DECODING TIRE ALPHA-NUMERICS

Tires are divided not only by brand, but by size as well. And this isn't as simple as it should be either—not just a 15" or 16" proposition. Instead, the numbers stamped on the side of each tire give a very precise assessment of that tire's physical characteristics: their height, width, speed rating, brand and purpose, and so on. Just because they have tiny sidewalls and an interesting tread design—or just because they have tall sidewalls and very little tread design at all—doesn't tell you everything you may need to know.

For example, take your average Bridgestone Potenza RE72 225/50/VR15.

Bridgestone refers to the manufacturer of the tire, while the Potenza name refers to a high-performance line of tires produced by Bridgestone. RE72 refers to the actual model of Potenza—there are many of varying quality. Don't assume because one Potenza is good the others must be similar. Not the case.

225 is an actual measurement, in millimeters, of the width of the tire. This measurement is the first step in guessing at the size of the footprint the tire will have on the road. It does depend though, on the width of the wheel you're installing the tire on. Think elasticity. A 225 tire on a 6" wheel is going to be taller than the same tire stretched on a 7" wheel.

50 describes the actual height of the sidewall, given as a percentage of the width. Since it directly depends on the width, not all 50-series sidewalls are physically the same height. On our 225/50/VR15 example above, 50 claims that the sidewall is half, or 50%, of the width, around 112.5 mm, while a 185/50/14's actual sidewall height would be a shorter 92.5 mm.

The V is the speed rating of the tire, a measure of the maximum speed the manufacturer of the tire rates the tire "good for," while R tells you that the tire is radial, as opposed to some made otherwise back before MR2s hit the road in the 1980s.

The Goodyear Eagle F1 has carbon-fiber reinforced sidewalls that significantly improve transitional response and overall handling "sharpness." Remember though: The rubber in tires is a type of buffer between the road irregularities and your posterior. The less it absorbs, the more you feel, both good and bad. Photo courtesy Goodyear Tires.

Alignment

The Idea: Optimizing the tire's contact with the road under a variety of loaded conditions

Just as simple—investment-wise anyway—as a set of tires is the issue of alignment. Alignments are crucial because they determine how efficiently the tires interact with the traveling surface.

As always, be certain that your suspension is OEM healthy before you think about a precise alignment. If you've got one component that is worn or out of adjustment and unable to function properly, it will be nearly impossible to establish any real knowledge of what it is your MR2 needs, and any adjustments you do end up making may yield compromised or coincidental performance gains.

Alignment is arguably the most underrated and ignored facet of any suspension setup. The 1991–1992 MR2s were especially known for their tendency to spin when haphazardly pushed past their cornering limit. This is not necessarily because of design flaws, but simple physics. Short wheelbase, mid-engine, rear-drive cars like the MR2 do not like any errors in input during at-limit cornering.

During these dynamic maneuvers, a driver input error such as closing of the throttle as the rear suspension is loaded under hard cornering can cause a sudden toeing-out of the rear suspension, resulting in the ends swapping places.

To the end of gaining some of additional handling stability, hopefully without sacrificing any of the sharp reflexes the MR2 is known for, there thankfully is some tuning available within the alignment.

Note: Alignment considerations should not strictly involve responsiveness or outright grip. Handling stability, tire wear, and tracking, especially on roads with distinct crowning, must be considered.

Balljoints and Tie-Rod Ends—Before making any alignment and general suspension modifications—especially on MR2s, and even more crucially on AW11 MR2s—you need to check the balljoints and tie-rod ends.

Shocks represent the most often-considered wear item on the suspension, but the less visible balljoints and tie-rod ends also experience consistent wear, and their condition can have a significant impact on feel, performance and overall safety of your MR2. Worn balljoints and tie-rod ends can also cause inconsistent alignment (resulting in poor handling performance and uneven tire wear), and intensified wear on the steering rack.

BASIC MR2 PERFORMANCE MODIFICATIONS & THEORY

Your performance alignment's main goal is to provide desired handling, braking and acceleration characteristics through optimal tire contact through a variety of suspension loads and overall dynamic transitions. This is different than what you want a commuter, which is straight tracking and even tire wear.

This AW11's alignment is set fairly neutral, with both toe and camber close to zero, resulting in a loaded tire alignment pretty much straight up and down, and toe set straight ahead.

There are checks the factory service manual recommends for establishing the condition of these items. If there is any question at all about their mechanical integrity, especially on AW11 and SW2x MR2s, replace them before slapping on that set of coil-overs. And replacing many of these wear items at once—balljoints, tie-rod ends, and various suspension bushings—can save you time and labor in the long run.

A High-Performing Alignment—For the non-enthusiast, alignments are usually only discussed when you install new tires, and then only for the sake of even tire wear, or when something is wrong, the most common scenario being a car pulling to one side or another while you're trying to drive straight. For you, the enthusiast, alignment is more complicated—and infinitely more useful as well.

When you begin to involve high-performance considerations, simply wearing your tire evenly, or being sure your car tracks straight, is insufficient. When accelerating, braking, turning left and right, trail-braking, and any other dynamic movement your MR2 sees, the tires are bearing the brunt of the physics of it all, doing their best to answer the steering wheel and drive wheels' request for direction (lateral) or motion (accelerating or decelerating). In these conditions of daily use, tires see unique loading that attempts to distort their natural position (whatever the alignment dictates when the car is sitting still). Aligning the tires so that they are most often at a position that provides the tire performance that you require is what we're after.

The three core components of a tire's alignment are toe, camber, and caster. Quantifying these measurements is done in degrees or fractions of an inch. Ideal alignment settings for drag racing, autocross, and road racing care often mutually exclusive. In fact, a well set-up car—and an adequately aware driver—are sensitive enough to alignment settings that handling on certain road courses or even autocross designs benefits from significantly unique settings.

Toe—Toe is a measure of the vertical inclination of a pair of wheels slightly toward or slightly away from each other, as seen from the front of the vehicle. If you are standing at the front of the car, toe-out would be trying to expose the inside of the wheels to you, while toe would attempt to "hide" the inside of the wheels. Toe can also be thought of as degrees of "pigeon-toe."

Toe-in enhances straight-line stability and overall, makes the car feel more planted and stable. This is good for the daily driver or commuting MR2. But toe-in also makes the car more resistant to direction change, and MR2s, while able daily drivers, are far more at-home performance driving. Everything's a compromise.

Toe settings front and rear generally have similar

Toyota MR2 Performance

Toe is a bit harder to see than camber. The rear toe settings on David Vespremi's SW2x are not as transparent as the negative front camber, which causes the front tires to clearly lean in at the top. This camber is especially noticeable when you compare the front tires to the rear. Photo courtesy David Vespremi.

effects, though for front-drive cars a bit of toe-out in the back can help get the car to rotate, and most cars generally more tolerant of front toe-out than rear.

MR2s are particularly sensitive to toe settings, especially at the rear.

So we know that toe-in dulls handling but increases stability, while toe-out increases the tire's tendency to change directions, but reduces overall stability. Toe-out increases as slight as 1/16" per side, depending on your MR2's other relevant specifications, can make it too willing to change direction.

For the commuting MR2, toe-out will require consistent correction from the steering wheel. Not only can this make your MR2 wander while driving on a daily basis, it will also increase the tendency to oversteer during high speed cornering, from off-ramps to race tracks.

Remember, start with an overall suspension setup where you don't have any single area—shock settings, sway bar setting, spring rate, etc., set to its limits—then experiment with alignment and tire pressure to get your MR2 reacting the way you want it to react. Modifications should be adjustments, not blind changes.

Camber—Camber is a measure of the "lean" of the tops of tires, either towards the center of the vehicle with negative camber, or away from the center if the camber is positive.

Negative camber allows the tires to maintain optimal surface contact when the suspension is loaded in a corner; without that negative camber, the tire would begin to lean over in a corner and ride on its edge, reducing the contact patch size and overall available grip.

Too much negative camber can reduce straight-line stability, as well as grip for acceleration and braking. Without the load of a corner, negative-camber tires are leaning "for no reason," and in so doing giving up some of that all-important contact patch. About 1 to 1.5 degrees of negative camber in the front is about the most you'll want to run on the street; more will sacrifice stability and tire wear. On the track, you can experiment with camber at negative 3 degrees and higher.

Note: Extreme toe and camber settings will radically increase tire wear.

Caster—Caster is the angle, pitched forward or backward, that the steering axis sits deviated from vertical, straight up and down being zero degrees of caster, negative caster being pitched forward, and positive caster being pitched rearwards. Visually, imagine a line through the wheel's center leaning forward or back, like a "chopper" motorcycle. That is exaggerated caster.

Positive caster improves turn-in without sacrificing tire wear, or the grip available for braking and acceleration grip like negative camber can. In this way, caster adjustment is preferable to camber.

Caster has significant input on handling feel through its effect on steering effort. This is a good thing, as it is one of the few areas you can affect steering wheel feel.

Positive caster will also allow your MR2 to naturally track straight, and increase high-speed stability, reducing the tendency of the short-wheelbase MR2 to "wander." Beyond increased steering effort, there is little downside to positive caster. That said, caster changes aren't always as noticeable as camber or toe adjustments.

Note: Remember to experiment with alignment settings until you find what works for your style and your MR2. So many variables go into what makes one setting optimal for your car—driving style, suspension setup, tire compound, where you drive your MR2, etc.—that experimentation is critical to find what that sweet spot.

Strut Tower Bars

The Idea: Stiffening the chassis for sharper handling

Strut tower bars are simple devices that support and add stiffness to the chassis. A stiff chassis will not deflect under hard cornering, helping to

Basic MR2 Performance Modifications & Theory

REAR TOE AND THE MR2

As we've mentioned, toe-in at the rear can help reduce oversteer, especially critical on a rear-heavy car like the MR2. This occurs mainly due to the reduced likelihood of catastrophic toe-out of the rear tires when lifting the throttle mid-corner; the more the tires are toed-in to begin with, the more dramatic the movement necessary for them to be excessively toed out.

Under acceleration, both the front and rear wheels will toe-in some. MR2s accelerate exceptionally well out of corners, mainly due to their mid-engine, rear-drive design, and slight rear weight bias. Get too eager with the throttle, though, and have to lift the throttle mid-corner for any reason and the rear tires will immediately toe-out. If the car is at or near the handling limit, this toe-out in the rear will cause the rear tires to lose grip first, and likely make the car spin. In a nutshell, this circumstance is what has given the SW2x handling a bad name.

This flaw or trait is subjective, and is discussed several times throughout the book. Clearly though, the SW2x is fairly intolerant of driver mistakes near the cornering limit. This is true, and very difficult to argue. For now, simply understand that lift-throttle oversteer is an issue with the rear suspension toeing out mid-corner—and, it must be mentioned, is not a condition unique to MR2s. So what can we do to help reduce this tendency?

Adding some toe-in to the rear can help reduce the tendency to dramatically toe-out. Toyota's lengthening of the rear trailing-arms 1993 helped address these issues, but for all intents and purposes, adding toe-in to the rear alignment of 1991–1992 MR2s can noticeable increase handling stability. Changing out the rear suspension is a giant task, and buying a 1993–1995 MR2 may have other compromises.

While adding 1/8" of rear toe-in is a good start, your best energy is spent focusing on proper steering wheel and throttle inputs, and keeping 10/10ths performance on a race track where mistakes are more safely made. It must be noted that choosing an alignment that allows the handling balance that fits your driving style is fine, but attempting to "dummy-proof" the handling will significantly numb your MR2's reaction and overall handling performance.

The SW2x is no stranger to oversteer. Dramatic changes in rear toe—and good old-fashioned driver error—is usually to blame. Photo courtesy Sarah Mays.

To reduce dramatic changes in rear toe, Toyota lengthened the trailing arms in 1993, shown here on David Martin's 1993 SW20. It worked. In lieu of such a dramatic change—the entire rear crossmember must be exchanged—upgrading to 1993-specification wheels, with some added rear toe-in, should noticeably increase handling stability.

maintain the integrity of the alignment we worked so hard to obtain.

Braces like strut tower bars help here. How much they actually help is entirely relative to the stiffness of the chassis you're using them on. That is, on an AW11 hardtop, with its closed roof and fairly stiff chassis, it may not offer the noticeable improvement it might on a 1967 Mustang convertible, or a T-top SW2x for that matter.

With their low relative entry cost, there really is no downside to strut tower bars unless you run into problems with clearance involving aftermarket intercoolers or intercooler piping.

Adjustable Sway Bars
The Idea: Reducing body roll

Sway bars, or anti-roll bars, are torsional mechanisms that resist the car body's tendency to sway from side to side in relation to the axles and wheels. Sway bars are attached to suspension points by much shorter, thinner bars called end-links. The sway bars are then anchored to the car's chassis, effectively attached to both the wheels and the chassis of the car itself.

When the movement of the wheels is relative to one another, loaded on one side in a turn, the sway bar resists that compression as function of its overall

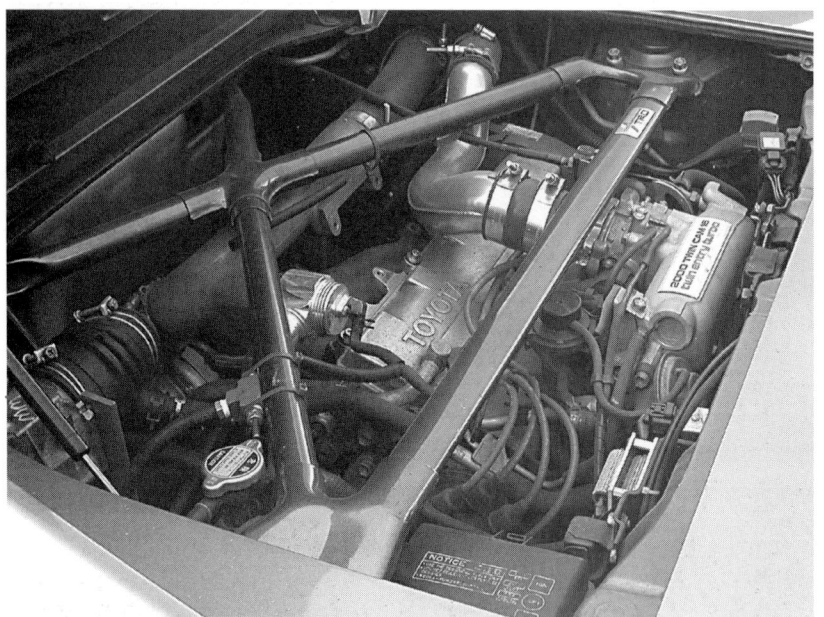

TOYOTA MR2 PERFORMANCE

Both the stock and TRD rear strut tower bars for the SW2x do the right thing by connecting both strut tops to one another, and then tying again to the firewall for additional bracing.

When suspension movement calls for both wheels to move—say, hitting a bump—the sway bar is allowed to twist within these rubber bushings. If these bushings are in bad shape, they can allow too much movement of the sway bar, reducing its effectiveness.

Sway bars are torsional devices that help ensure ideal tire alignment with the traveling surface by mechanically resisting roll. On this SW20 crossmember that has been detached from the car, you can clearly see the difference between the sway bar, which resists roll, and the crossmember, which provides physical structure for the rear suspension.

size and thickness, as if a stiffer suspension spring has been installed.

Remember, wheel/tire alignment is critical. We work hard to keep our tires ideally aligned, and when your MR2 tries to lean over on its side in a turn, undesirable alignment like positive camber can occur, drastically reducing grip. Sway bars can help reduce this tendency. The neat thing about sway bars is that they give so much, and ask for little back in return.

We've discussed the idea of compromise, and understanding what you're giving away to get what you're hoping for, but sway bars don't require much compromise. Unlike 400-lb. suspension springs, or that booming 3" exhaust, sway bars quietly reduce body roll with just a slight impact on ride quality.

Note: By transferring more of the load from the front contact patches back to the rear, a large front sway bar can improve traction (to the rear wheels, of course) out of corners, acting very much like a limited slip differential in that regard.

Choosing a New Sway Bar—People often seek out a certain sway bar thickness—the thicker the bar, the more resistant it is to roll—but the issue of the difference between the thickness of the front, and the thickness of the rear should not be forgotten, especially as they collectively relate to the stock setup. For example, take your impressions of your stock sway bar setup and your assessment of the handling balance they provided.

As an example, on the SW20 the TRD (Toyota Racing Development, Toyota's in-house performance aftermarket and racing division) aftermarket front sway bar is 38% stiffer than stock, while the rear is only 22% stiffer than stock. This implies that the stock setup would provide a setup more prone to oversteer than the TRD bars, though of course, as with most participants in the black magic of suspension tuning, it isn't that simple. The moral here, though, is easier to understand: when choosing sway bars, don't just consider thickness—give more attention to the ratio of bias front to rear.

Also consider your application: controllable

Basic MR2 Performance Modifications & Theory

A TRD white rear sway bar is clearly visible on this SW20—the white bar running perpendicular to the exhaust. The TRD rear sway bar is 22% stiffer than the stock rear sway bar.

CALCULATING SWAY BAR STIFFNESS

So you want to know how much stiffer that 24mm ST front sway bar is than your stock 19 mm bar?

The stiffness of a set of sway bars can be calculated from its diameter, being a function of its fourth factor. For example, our stock SW20 19mm front sway bar offers a stiffness of 19 x 19 x 19 x 19, in this case 130, 321. The 24mm ST front sway bar would calculated the same way—24 x 24 x 24 x 24, which gives us 331,776 units. Dividing the latter by the former—331,776/130,321—shows us that the ST bar is more than twice as stiff, 2.55 times! Very minute changes in sway bar thickness, then, yield large differences in stiffness.

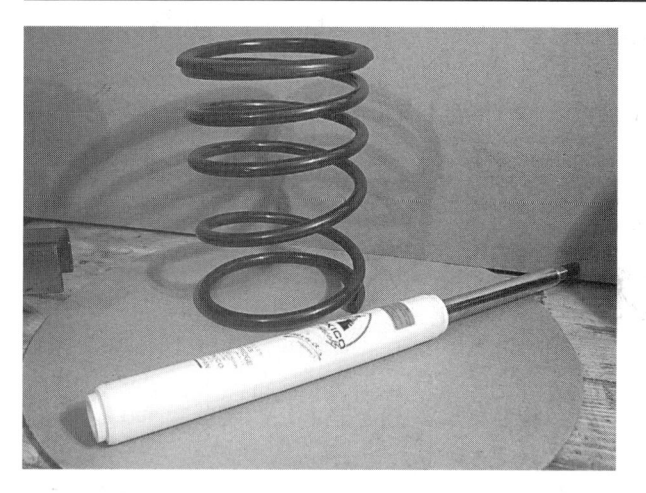

Suspension springs significantly impact the ride quality of your MR2, perhaps more so than any other suspension component. Pictured are Eibach Pro-Kit progressive springs for the SW2x, paired with 5-way adjustable Tokico Illumina shocks.

oversteer works far better on an autocross course than it might on a road course. As speeds increase, so does the speed in which that oversteer comes on and sweeps you right off course. Ultimately, the ideal setup is a set of adjustable sway bars like that available from Suspension Techniques that will provide an adjustable range of bias. For this reason, the sway bars from Suspension Techniques come highly recommended. Consider individual setup preference as well. More than likely your stock sway bar produces understeer, but how badly? What is your driving style? If you like to hang the tail out and drift, you'll want "more" rear sway bar than front, i.e., alter the thickness bias from stock between the front and the rear more towards the rear. If you prefer balance, you'll want to do that less, and stay closer than the ratio established by the factory.

And always consider adjustable sway bars first. While adjustability in suspension application can be a double-edged sword, don't paint yourself in a corner by purchasing non-adjustable sway bars unless you are absolutely certain that a set offers the roll stiffness and ratio of bias front to rear that you demand.

One note of caution: Stiffer sway bars often require reinforcement of the sway bar mounting brackets. The stock brackets were manufactured assuming stock sway bars, and are apt to fail as you start significantly increasing the size of the sway bar. All that twist-resistance energy has to go somewhere.

Note: Unlike springs and shocks, sway bars will not decrease the pitch or squat of the car as it accelerates and brakes—movements that do indeed waste energy and time.

Springs

The Idea: Controlling the movement—laterally, fore and aft—of your MR2

Suspension springs support the weight of the car, and dissipate its movement while the car is in motion through energy transfer—compression and decompression. Stiffer, shorter suspension springs will allow the car to transfer weight less dramatically, and lower the car's center of gravity, improving handling basically by wasting less body movement.

Linear vs. Progressive Springs—Springs are quantified by their ability to resist compression, and are so described in pounds, a 450-lb. spring being stiffer than a 350-lb. spring (such a spring requiring a 350-lb. load to compress the spring 1"). There are two types of springs, linear and progressive.

Linear springs exhibit the same resistance throughout their range of compression, and are typically thought of as less streetable due to this seemingly uncompromising nature. They are also more trusted by many drivers, with a single spring rate easier to feel and adjust to than one that changes.

Toyota MR2 Performance

This SW2x has stock suspension springs from a 1996 JDM MR2 GT-S installed. These springs were a compromise between the factory springs, and much stiffer aftermarket springs. Roll stiffness was slightly increased without excessively reducing suspension travel.

While the stances looks good, shorter suspension springs can also reduce the suspension travel, which can become a serious issue when the suspension is fully loaded, in a turn or over a bump. This 1994 Turbo is lowered with coil-overs—Sustec Pro units by Tanabe. Photo courtesy Brian Hill.

In addition to performance and ride comfort, spring height also has a lot to say about the driveability of your MR2. Speed bumps, awkward lot and driveway entrances go from thoughtless to a chore. Photo courtesy Jensen Lum.

Progressive springs are designed to offer less resistance through the initial stages of its compression in hopes of providing a more supple ride for everyday bumps and potholes, while becoming increasingly difficult to compress as suspension loads increase. Progressive springs have their "weight" given as a range for this reason.

Suspension Travel—Suspension travel is the potential range of movement or motion of the suspension. It is usually brought up in discussing the travel of the dampening device—the shock—courtesy of a shorter suspension spring. Proper suspension travel—that is, a shock and spring combination physically able to fully and freely compress and decompress throughout its intended range of motion—is critical for the suspension to do its job. Especially considering the MacPherson suspension design the MR2 is forced to work with, if the springs are too low, or so stiff that they are poorly matched to the dampening ability of the shocks, the handling will end up noticeably worse than stock.

A huge problem is springs with a spring rate that is too low and drop the car too low. Given that the shocks have a certain distance they can travel before they bottom-out on the bump stops, it makes sense that lowering springs, through simply being shorter than the stock springs, will use some of that shock travel sitting still, entirely uncompressed. Because of this, any lowering spring must be stiffer to resist full compression; as a byproduct of being shorter, they simply have less room left to move.

For street-driven MR2s, try to keep your lowering level 1" or less. While greater drops might be workable in certain track-oriented setups, and will undoubtedly look aggressive, in those dedicated situations a full coil-over suspension is a better choice. An extremely low spring, coupled with an aftermarket shock, is a crude combination by comparison.

Once lowered, caution must be used when driving on public roads. Every driveway and speed bump becomes an event, something people following you too closely generally aren't sympathetic about. On real-world roads, with bumps, expansion joints and road irregularities, a suspension with adequate travel and compliance will serve your MR2 far better than a low, stiff setup, and is probably faster from point A to point B too, no matter how much body roll and suspension movement you might feel.

Body roll, while alignment-wrecking and best reduced, cannot be entirely corrected with a

BASIC MR2 PERFORMANCE MODIFICATIONS & THEORY

suspension that doesn't travel. Body roll can certainly be cleaned up though, and is best done by a set of adjustable sway bars, while still allowing for appropriate range of travel for your suspension.

Spring height and rate must be carefully paired as well. Stiff lowering springs will be much harder on the stock shock than the stock springs those struts were designed for. (In fact, many springs are manufactured for use with the stock shocks and their soft dampening in mind, and oftentimes this fact isn't made obvious by the spring manufacturer. Obviously a spring can't be ideal for both stock and non-stock shocks, so be sure you purchase what you need.)

Lowering the car with aftermarket springs while keeping the stock shocks (which probably have their share of miles on them already) is not the way to make a fast-handling MR2, and is not a process I recommend. If you only can afford one modification—shocks or springs—get the shocks. If you can afford both shocks and springs, buy quality adjustable shocks and save the balance for a set of adjustable sway bars. If you can afford both quality adjustable shocks, springs, and sway bars, consider coil-overs in place of the suspension springs.

Shocks
The Idea: Better spring control
Note: The MR2 has always used a MacPherson strut suspension to some degree, either in the front, or both the front and rear, technically strut cartridges containing inserts. For ease of communication though, the term "shock" will be used interchangeably with "strut," "strut insert," etc.

It is the shocks' job to control the suspension that would otherwise move and bounce freely. The springs, bushings and sway bars all move in reaction to your MR2's motion, whether it is a speed bump in a parking lot, or an off-camber turn on a road course. Unlike lower, stiffer aftermarket springs, which ideally necessitate upgraded shocks, stock and aftermarket springs alike can benefit from a stronger dampening shock.

As always, a compromise must be reached: too soft a shock allows too much suspension movement, while a shock that is too stiff can ruin available traction by not allowing the suspension springs to release the energy of bumps and other road irregularities, not to mention deliver a harsh ride.

Like a pogo stick, a shock has two methods of movement: compression and rebound. The compression comes into play when the suspension is loaded with weight of some kind, whether physical weight, while the car is sitting still, or

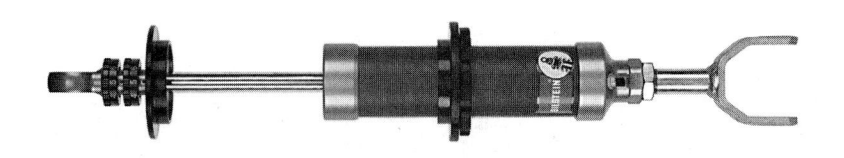

Tokico, Koni and Bilstein all make consistently good shocks for a variety of applications and budgets. This shock by Bilstein features their Modular Damper System and is a "two-way" shock, meaning it is adjustable for both compression and rebound. Photo courtesy Bilstein.

dynamic weight when the suspension compresses in a turn, braking or accelerating. The shock helps to reduce the springs' desire to continue to bounce upon their decompression. Shocks that are adjustable in both directions—for both compression and rebound—allow further fine-tuning of the suspension for application, but are more expensive.

Be careful with your spending here. Koni Yellows and Bilstein shocks are extremely high quality but their cost, coupled with a quality aftermarket spring (if you planned on going that route) can quickly approach or exceed the cost of a coil-over setup. If you're sticking with stock suspension springs, a stock class autocrosser for example, this is not an issue, and a set of Koni yellows is indispensable.

BASIC BRAKING PERFORMANCE
Brake Upgrades
The Idea: Increased fade resistance, shorter stopping distances

Defining Braking Performance—Like the vague term "handling," what exactly is meant by "braking performance?" Braking performance is a broad term that can be summarized as:
- stopping distance
- fade resistance
- pedal feel
- brake balance/bias front to rear
- sheer caliper/"clamping" power

Friction between the brake pad and brake rotor, and between the tire and the road surface, are also contributing factors to braking performance. Any weakness in these systems will reduce performance.

Better Brake Pads, Better Brake Fluid—Special brake fluid and pads designed to operate effectively at much higher temperatures are available from a

Toyota MR2 Performance

Larger brakes increase the surface area of contact with the brake pad—the swept area. Here, a stock SW20 rotor is compared to an aftermarket two-piece rotor. A benefit of the two-piece rotor design is the ability to increase rotor size without dramatically increasing weight. Photo courtesy Jensen Lum.

The same two-piece rotors installed on an SW20 with beautiful Volk wheels only slightly obscuring the beefy Wilwood calipers. This is an extreme brake setup for users demanding lightweight braking components with a larger heat-sink capacity than stock. Photo courtesy Jensen Lum.

variety of manufacturers, and offer significant increases in overall braking performance. They simply deal with heat better—and coupled with high-performance tires, for the vast majority of MR2s this is all you'll ever need in braking performance. It takes an extremely punishing course, a huge amount of horsepower, or a very "braking-aggressive" driving style to push a "Stage I" braking system like this past its limits.

Supplemented by a quality pad compound and brake fluid with a higher (than stock) boiling point—perhaps some homemade brake ducts for airflow around the brakes if you're feeling industrious—this setup is wonderful not only for its performance, but its simplicity, quality, and inexpensiveness. There is something to be said for being able to use as many OEM parts as possible, until you are absolutely sure their replacement will consistently provide more aggregate gain than loss. Until you've performed these simple modifications and had fade issues with that setup, replacing your braking system is generally not necessary, and can be a tremendous waste of resources.

Brake Fluid

Brake fluid is rated by its ability to resist boiling, either "wet" or "dry." Wet boiling point is the temperature at which the brake fluid boils when moisture is present, while dry boiling point is the boiling point of the fluid with no water at all. In a perfect world, all brake fluid would be installed completely dry and stay that way, but doesn't happen in the real world.

Brake fluid provides the hydraulic pressure to operate the brake system, and lubricates the internal brake system components, providing corrosion

---

EVALUATING A BRAKE PAD

Brake pad composition is sort of like alignment or tire pressure settings—somewhat underappreciated for its performance potential, and more complicated than it seems.

Most crucially, aftermarket brake pads are made of some material better able to withstand higher temperatures than lesser pads. Additionally, performance brake pads must also offer increased friction between itself and the rotor. Unfortunately, all aftermarket pads are not equal, and a variety of considerations must be considered when looking at brake pads. Brake pads are usually characterized by their friction, fade resistance, cold performance, cost, noise and dust

Brake Pad Types—The extremely high temperatures absorbed by brake pads are enough to cause many non-reinforced pad compounds to lose structural integrity, and so brake pads are reinforced by the application of certain metals and fibers, from glass fibers and and copper, to more exotic ceramics.

Brake pads available for your MR2 will be broken down into three basic categories.
• Metallic: reinforced through metallic based (steel, copper, etc.) supplements
• Semi-metallic: reinforced through a combination of metal and organic supplements
• Organic: reinforced through mainly organic, non-metal supplements

Generally speaking, organic brake pads provide less structural strength, but also promote extended rotor life. The more metal embedded within the pad, the stronger the pad—and the more "metal on metal" each braking procedure will endure, increasing strength but reducing rotor life.

Basic MR2 Performance Modifications & Theory

Aftermarket rotors like the EBC 3GD rotor utilize advanced casting technology in an attempt to safely alter a standard blank rotor—drilling and slotting it, for example. Photo courtesy EBC Brakes.

BRAKE FLUID DOT LEVELS

DOT Level	Dry	Wet
DOT 3	401°F	284°F
DOT 4	446°F	311°F
DOT 5	500°F	356°F
DOT 5.1	518°F	375°F

Popular Aftermarket Fluids

	Dry	Wet
AP Racing Formula 5.1	527°F	363°F
AP Racing 551	527°F	302°F
AP Racing 600	527°F	410°F
ATE Super Blue	536°F	396°F
ATE Typ 200	536°F	392°F
Castrol LMA	446°F	311°F
Castrol SRF	593°F	518°F
Ford Heavy Duty DOT 3	550°F	284°F
Motul RBF 600	593°F	420°F
Wilwood 600+	600°F	417°F

protection as well. It is a fluid's performance in these areas, along with its resistance to wet and dry boiling, that decides its DOT level. Generally, the higher the DOT level, the higher the performance of the fluid.

Evaluating Brake Fluids—So a quality fluid is priceless for repeatable, aggressive braking, and such fluid is broken down by its DOT level, and further by its specific boiling points. What else do we need to know?

Though boiling point is obviously critical, how well a brake fluid resists moisture absorption is crucial as well. Whether through storage, flushing, or some other happenstance, any moisture allowed to come in contact with premium brake fluids will drastically reduce its boiling point, and thus its performance. Motul RBF 600, for example, is very hygroscopic—that is, it has a high affinity for water, and should only be used for racing and changed regularly.

As we can see from the difference between wet and dry boiling points in the chart, even a small amount of water drastically reduces the boiling point of the fluid. So for high-performance use, it is best to change aftermarket brake fluids more often, which makes the more expensive fluids less than ideal for street use.

Buying an expensive brake fluid that has a higher tendency to absorb water, and that you're going to change every 18 months, doesn't make much sense. The bottom line: if you're going to change your fluid regularly, you can go with a more complex fluid with a higher dry boiling point; if you're not—and be honest with yourself—use a fluid with a higher wet boiling point.

Note: Do not mix brake fluids! Mixing brake fluids can compromise that brake fluid's performance, leaving you with an uncertain mixture of fluids that collectively perform less than they might've alone.

The Cons of Big Brake Upgrades

As far as actual clamping power, balance and pedal feel, mechanically this is decided by the design of the caliper (namely size/number of pistons), rotor size, and a complicated tangle of hydraulic pressure and friction of materials. Most factory brakes, including each generation of MR2, use one or two-piston setups in the front, and one-piston setups in the back, while most aftermarket braking setups will up the piston count up front to four or even six, and two in the rear (if the upgrade addresses the rear at all). A properly designed aftermarket braking system will improve pedal feel

EBC separates the performance of their brake pads in an easy-to-understand color scheme—Greenstuff, Redstuff, and Yellowstuff, and Bluestuff. Their pads also use a "brake-in" abrasive coating to assist breaking in a new brake-pad-and-rotor setup. Photo courtesy EBC Brakes.

and modulation, and can improve balance as well. So if larger brakes can absorb more heat, increase clamping power, and offer improved pedal feel, why shouldn't you get them?

For one, MR2s are blessed with excellent braking performance from the factory, largely in part to their relative lightweight and mid-engine design. If they were not, supplementing factory performance would be more critical. Since we think of modifications as adjustments though, and factory MR2 brakes require less of this adjustment than many other performance cars, brakes are not a crucial area for most MR2 owners, whereas handling stability and powerband might be.

As we've mentioned, very basic upgrades can go a very long way in providing fade-free, short stopping distances. Keeping things simple and OEM until you absolutely must have some adjustment of their stock performance is highly advisable.

Second, a quality big-brake setup costs big money. As an example, the kit from GREX for the SW20 costs almost $2000 for the front brakes alone, and ZZW30 kits are similarly priced; for the AW11, there is almost no aftermarket support at all for big brakes. $2000 will buy a lot of suspension work, or, better yet, a lot of track time. Remember, keep it simple, and only modify based on an identified need for adjustment, not a perceived need (marketing-induced desire).

Finally, big brakes are also usually very heavy unless you get into some very expensive setups. They are also often universal applications modified to fit your car, often with inherent design or construction compromises which can reduce the likelihood of quality service on your MR2s for years and years like your OEM setup provided.

Additionally, a big brake kit will often demand larger wheels to clear the larger calipers, upping once again the cost and complexity if your current wheels won't clear.

Note: Stainless steel brake lines are not crucial upgrades, but can improve pedal feel and braking modulation, especially if your MR2 has some miles and age. The worse condition your current brake lines are in—they generally suffer with age—the greater potential reward of this modification.

BASIC ENGINE PERFORMANCE UPGRADES
Exhaust System
The Idea: To evacuate the exhaust gases out of the cylinder head as efficiently as possible

Probably the simplest and most common of all automotive performance modifications, a properly designed and implemented exhaust system can offer at least some sort of performance gains to almost any setup, while at the same time shouting to the world that you are no longer stock. Note that though exhaust work can impact the performance of any car, it is far more critical on forced-induction cars like the SW20 than it is on the AW15.

For the AW15, the general recommendation is an OEM exhaust for most applications, and for a slight performance increase, an OEM AW16 muffler. Once you have made enough modifications to significantly impact the volumetric efficiency of the engine, larger exhaust sizes can be beneficial, but for the vast majority of AW15s out there, stock AW15 or AW16 mufflers with quality, mandrel-bent piping is sufficient. On any forced-induction car, which this section mainly addresses, consider a large, free-flowing exhaust system a necessity.

Evaluating an Exhaust System—So what constitutes an exhaust system? Generally considered, the exhaust system includes everything from the exhaust manifold to the tip of the muffler. This includes any resonators, flex sections, downpipes, catalytic converters, clamps, and hangars. The exhaust needs of each generation and model of MR2 are covered in their respective staged buildups.

A stock downpipe and primary catalytic converter on an SW20 will limit how much power that shiny new 3" exhaust muffler and muffler piping unleashes, while a new downpipe's power production depends on the nature of the setup downstream from it, namely the ID (inside diameter) of the piping and how free-flowing the muffler is. In this way, the collection of connected, cooperating parts from the exhaust manifold to the

Due to its unique exhaust routing because of its mid-engine layout, and the resulting nature of its rear-bumper design, the SW20 shows off exhaust systems exceptionally well. Photo courtesy Michael Tedone.

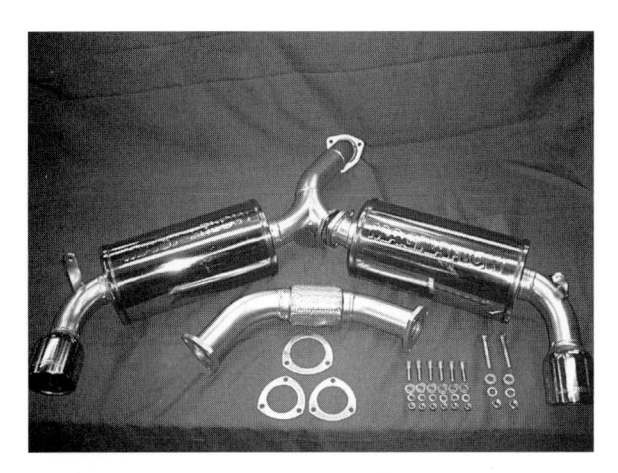

Basic MR2 Performance Modifications & Theory

An exhaust system technically begins at the exhaust manifold and ends at the tip of the muffler. The quality of design and construction between those points can have a significant effect on power output and sound. KO Racing is among a number of companies manufacturing quality exhaust systems, in their case in very limited numbers. Photo courtesy KO Racing.

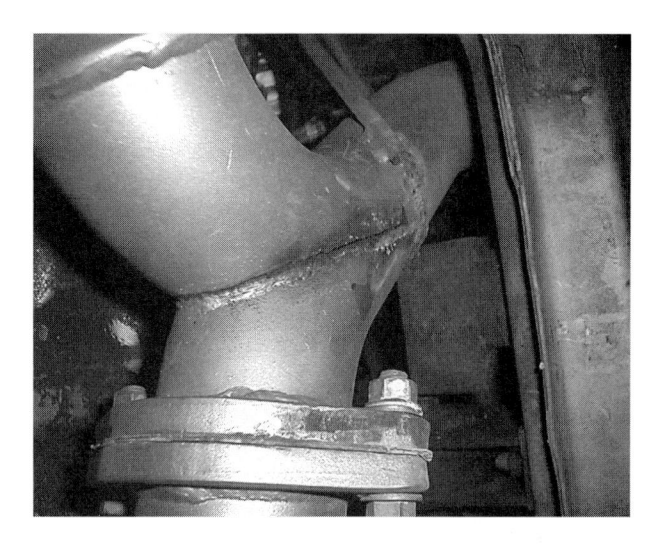

Why insist on quality stainless steel materials, clean beads and a flex pipe on any exhaust system, custom or aftermarket? Consider what would happen if this very expensive aftermarket exhaust should suffer a catastrophic material failure. Note the crack!

Welds and quality of materials are critical players in the resulting quality of an exhaust system, and are unfortunately overlooked by many enthusiasts who are instead merely considering aesthetics or dyno-charts. The weld beads on this exhaust from KO Racing are neat and clean. Photo courtesy KO Racing.

muffler tip is very much a system.

Note: As a basic rule of thumb, on a turbo-charged car you want the largest exhaust that provides the noise level you can live with every day: simple, straight-through and made of stainless steel (though titanium is nice if you can afford it). If yours is a daily driven MR2, go with one noise-level quieter than you think is ideal.

On a normally aspirated MR2, a quality aftermarket exhaust should feature mandrel bends, a stainless muffler and an ID (inside diameter) appropriate to your overall setup. Regarding the prospect of backpressure, and the argument whether or not backpressure is helpful on normally aspirated applications Stephen Gunter says, "On any engine, I want the least amount of back pressure possible from the exhaust system. I'm much more concerned with pipe diameters and lengths, and there are mathematical formulas that assist with proper (exhaust) system design. A complete exhaust system has to be designed with all engine factors in mind, including individual cylinder displacement and intended rpm range of use."

Headers

4A-Gx, 5S-FE, 1ZZ-FE—In general, header modifications on normally aspirated MR2 engines are unnecessary. One of the very few available aftermarket exhaust manifolds for the 4A-GE is a header made by TRD. The TRD header for the 4A-GE has proven to offer minimal—if any—horsepower gains on stock to moderate 4A-GE builds, and is more of an expensive method of saving weight, the header weighing less than the OEM exhaust manifold. The TRD header on a 4A-GZE is more worthy of your consideration considering the 4A-GZE is supercharged, and thus is more sensitive to exhaust flow restrictions, though even this modification only really becomes important as factory boost levels are surpassed. The factory exhaust manifold on either 4A-Gx engine flows exceptionally well, and for most AW11 owners your money would be better placed elsewhere in the car. A quality aftermarket header, coated, with a new OEM exhaust manifold gasket and installation, could end up costing $600 or more, and this is assuming the rest of your exhaust system will not require modification to bolt to the header. Or, worse, if your exhaust system is still stock, any gains at all will be snuffed out by that stock exhaust. Adding the cost of exhaust work, a header and exhaust combination could end up running more than $1000—enough to put you well on your way to a 4A-GZE swap, or even another used MR2. Think about it.

The 5S-FE will benefit slightly with a quality header, but file this modification right next to camshaft upgrades for the 5S-FE as well: not enough bang-for-the-buck for most folks. Remember, any change you make to the car should be an adjustment, and $500 later after installation, new gaskets, and the like, for most owners a header is simply not going to adjust enough to warrant its cost, and loss of factory fitment on most normally aspirated MR2s.

Turbocharging a 1ZZ-FE, you obviously have little choice but to replace the exhaust manifold, while the other assortments of 4A-GEs, 20-valve to small-port 4A-GZEs all depend on what you've got and what you're trying to do with it. The factory 20-valve exhaust manifold is tubular, and assuming your modifications to the engine are limited, swapping it out for an aftermarket manifold won't do you much good.

SW20—As is the case with its normally aspirated brothers, the turbocharged SW20's factory exhaust manifold also presents no significant restriction, and so replacing it is not only unnecessary, but can also be troublesome.

A turbocharger is very heavy and places a load on the end of the exhaust manifold where it is attached. The short, thick-walled stock exhaust manifold holds up well to this weight, but aftermarket tubular exhaust manifolds are not only prone to cracking, they also stress the hardware that anchors the manifold to the cylinder head, making annoying exhaust manifold leaks more common.

If you have found a tubular exhaust manifold for your 3S-GTE that you simply must try, be sure that any kit you buy has addressed the issue of cracks and exhaust leaks with significant bracing, and robust manifold design.

SW20 Exhaust System Components

The SW20 factory exhaust looks like a dual exhaust, but a dual exhaust on a 4-cylinder engine isn't very likely. The SW20 is simply a muffler turned parallel to the shape and direction of the trunk, with tips protruding from either side. Factory exhaust systems are mass-produced, low-

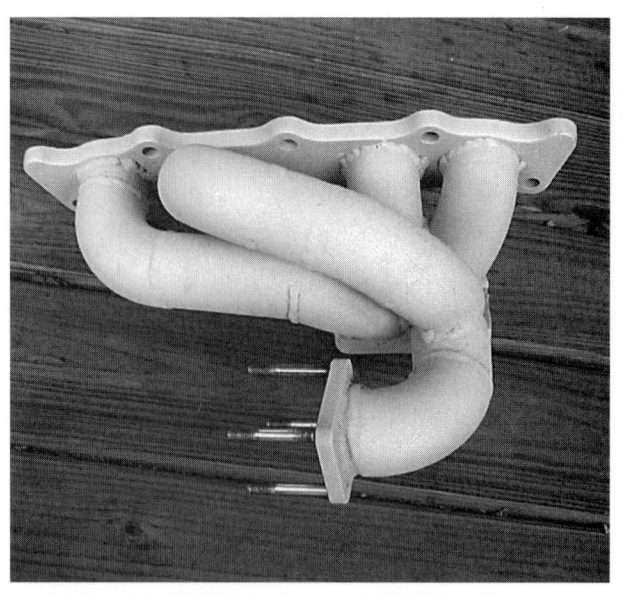

This aftermarket, one-off, 3S-GTE manifold features sturdy, robust construction to help resist cracking. Photo courtesy Jensen Lum.

buck items, and are thus very heavy.

Two stock catalytic converters are plumbed into the stock SW20 exhaust system: one big, bell-mouthed unit right off the turbo, and another at the B-pipe that crosses across the back of the car, parallel to the stock muffler itself. The B-pipe contains a secondary catalytic converter on the factory SW20 exhaust system, and is very restrictive. Replacing it with a straight pipe, or simply a stainless pipe with no catalytic converter, will increase your turbocharger's ability to spool, and should allow the car to make 1–2 additional psi without a boost controller simply due to the decreased exhaust backpressure. On a stock SW20, this modification is worth 8–15 whp.

Aftermarket downpipes replace the primary catalytic converter, allowing the turbocharger to spool up faster, and produce more torque throughout the powerband, while the secondary catalytic converter can be replaced by a straight-pipe, or test-pipe that removes the car and replaces it with a simple empty 2.5"–3" pipe (depending on the rest of the exhaust configuration). Removing these catalytic converters, which is legal only for off-highway/race track use, will significantly reduce exhaust backpressure, resulting in increased torque and horsepower. It will also allow more boost—even without a boost controller to manipulate boost settings. A downpipe and a full open exhaust can often increase boost enought to hit fuel cut, especially on cool mornings, and allow

Basic MR2 Performance Modifications & Theory

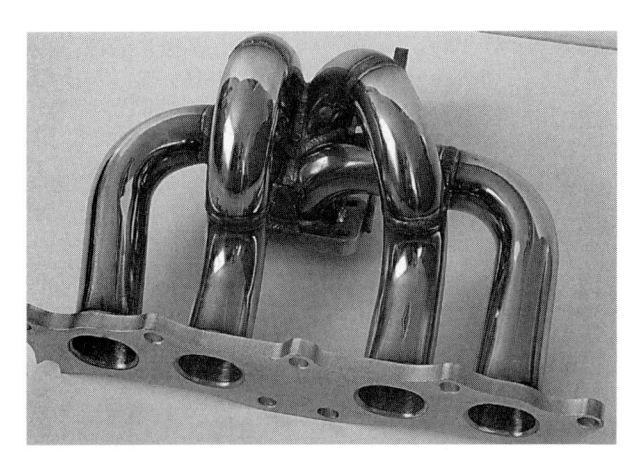

The construction of tubular exhaust manifolds must be robust—and braced, ideally. Heat and vibration can stress the construction and hardware of aftermarket manifolds, which must have something to do with most OEMs either only very selectively employing their use.

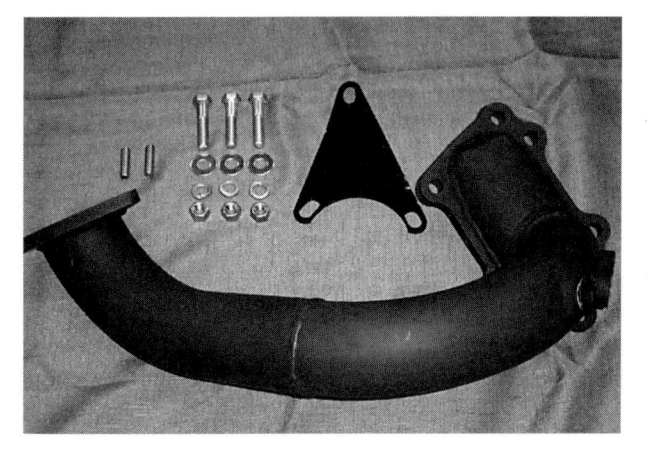

Another sign of quality is the thoroughness of the provided hardware. In this case, KO Racing has provided all the necessary hardware for installation of their aftermarket downpipe—including a triangle-shaped brace to support the downpipe and resist cracking. Photo courtesy KO Racing.

the turbocharger to spool more quickly to boot.

Exhaust Systems and Weight—Not only will removing both cats open up the exhaust flow out of your cylinder head considerably, but it will lighten the rear of your MR2 as well. Replacing the heavy stock muffler with a lightweight muffler will lighten the system as well, all three combined saving anywhere from 20–50 lbs., depending on the weight of the aftermarket components you replace them with. Combine this with a lightweight battery, and you can save 75+ lbs. without trying—and don't forget, the SW20 has a slight rear-weight bias, and removing weight from the back will help improve this.

Gutted Catalytic Converters—The aftermarket tuner Blitz used to make a defuser for the 3S-GTE, which was basically a gutted cat that said "Blitz" across the side. Simply gutting the catalytic converter is a similarly effective piece, free assuming you do the labor yourself, while a good downpipe is around $250–$300. Gutting the downpipe amounts to removing the honeycomb material that fills the converter housing. This is generally done with a hammer, chisel, block or wedge of some sort to remove all debris. Ideally though, gutting is a last resort, and a true downpipe is a safer, better all-around choice.

Gutting the cat also retains the factory "elbow"—a dramatic bend in the cat itself that is largely responsible for the restriction in airflow. A gutted catalytic converter is an inexpensive, but ultimately compromised solution. An additional side note in the gutted-cat vs. full-downpipe argument regards weight: a typical downpipe weighs about 8 lbs. less than a gutted cat. For these and other reasons, a true downpipe is the 3S-GTE owner's best solution.

Note: Interestingly enough, in the Gen III version of the 3S-GTE, Toyota did away with the "elbow" bend, one of its many improvements.

Exhaust Silencers—Many exhaust systems come with removable silencers—little metal corks that reduce noise levels (and horsepower). They help keep things quiet when you want to hear what your passenger has to say, and are easily removable, usually via a single bolt and a quick tug when noise levels are not a concern.

Though the silencer will limit horsepower gains by effectively restricting exhaust sizing at the tip, they are removable so that you can go back and forth between big power and quiet operation.

Ignition

The Idea: Increasing spark energy for increased efficiency and power

The 4A-GE, 4A-GZE, 5S-FE and 3S-GTE all use a traditional, distributor-based ignition system, while the 1ZZ-FE uses what Toyota calls S-TDI, Single Toyota Direct Ignition, which is distributorless. It is the job of each of these systems to provide the spark that lights the air/fuel mixture.

Obviously your MR2 comes with such a system stock, and for some cars there is improvement to be made here. The factory system may be merely adequate, and a more thorough ignition can increase the intensity of the explosion in the combustion chamber, and thus a more complete air/fuel mixture burn. This is usually not the case with MR2s of any generation, however.

For the 4A-GE, 4A-GZE, 5S-FE, 3S-GTE and

Toyota MR2 Performance

```
Customer        :                       Miles           : 0.0
License         :                       Weight          : 2830.0
VIN             :                       HP @ 50 MPH     : 13.00
Yr/Mk/Mdl       : 91 Toyota MR2         Cyl/Disp.       : 4/2.0
Comments        :
                        HorsePower Curve Test Results
Test Run        :                       Base Run        :
Max Power       : 251.6  @ 6950 RPM     Max Power       : 246.5  @ 5950 RPM
Max Torque      : 261.9  @ 3950 RPM     Max Torque      : 246.0  @ 3950 RPM
Comments        : KO Downpipe
```

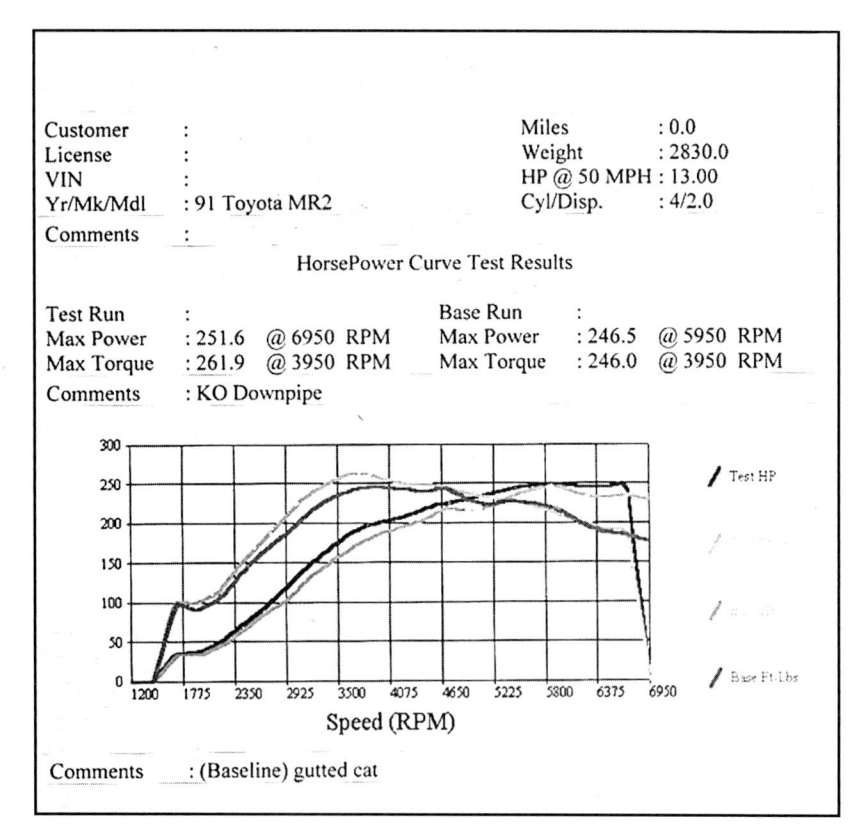

Comments : (Baseline) gutted cat

This dyno test by KO Racing demonstrates the performance of a true downpipe vs. a gutted cat. Eliminating the "elbow" on the stock cat made an extra 16 lbs-ft. of torque, and while making only an additional 5 hp, it also effectively widened the powerband to the left and right. Testing and image courtesy of KO Racing.

This engine ground kit was marketed for the Nissan SR20VE, but grounding kits are usually universal, and can easily be custom made, so if the length is right and the end-sizes are compatible, it could just as easily be used in your MR2. They are simply wires that provide a ground, nothing more.

1ZZ-FE, in almost every case OEM ignition components are ideal, meaning Genuine Toyota plug wires, distributor cap and rotor, without any ignition amplification of any sort. The 3S-GTE is especially sensitive to ignition components, and has a tendency to run poorly when you're using inferior components, either aftermarket or OEM. Ignition components should be replaced frequently, as they tend to fail in small, difficult-to-notice increments. Change copper plugs with the oil (see following section on 3S-GTE spark plugs, below), and factory ignition wires at least at factory recommended intervals.

If you're running boost levels beyond 20 psi, the resulting 350+ whp might indeed be taxing the factory ignition system to the point where an adjustment might be necessary. But until then, leave well enough alone. Provided your ignition system is OEM-healthy, leave the majority of your factory ignition system alone—with one exception: 3S-GTE spark plugs.

3S-GTE Spark Plugs—Copper spark plugs are commonly used in turbocharged applications for the superior conductive properties of copper.

Additionally, one of the most significant benefits to the copper plugs comes courtesy of one of its downfalls: rate of wear. Instead of the extended life enjoyed by the OEM platinum plugs, copper plugs should be replaced every 3000–5000 miles—basically with the oil.

Why would this be a good thing? A tuned, optimized turbocharged engine has a smaller tolerance for poor maintenance than a stock engine, and needs to be constantly monitored to ensure everything is doing what it should; piston rings should be within factory tolerances, as should turbocharger seals, the head gasket, and so on. Reading your spark plugs can offer signs of many sorts of internal failures, the aforementioned seals and gaskets notwithstanding. In this way, removing and replacing your copper plugs every 3000–5000 miles helps you keep tabs on what's going on inside your 3S-GTE.

Copper plugs are also very inexpensive, so their frequent replacement isn't a financial concern. Order a year's worth and store them next to the Mobil 1 Synthetic and Genuine Toyota oil filters in the garage.

Spark Plug Heat Range—As critical as metal type is heat range. Heat range is basically a kind of quantification of the spark plug's ability to dissipate heat from the combustion chamber. In a nutshell, you want a plug hot enough to completely ignite the mixture without fouling, but cold enough to prevent pre-ignition. The lower number given to a plug, the colder that plug is, so that a 3 would be colder than a 7. Tuners often run a colder plug for detonation resistance, again, only useful to a point.

Basic MR2 Performance Modifications & Theory

Spark plug wires like these from uber-rare Garage Fukui SPL/Phoenix Power are more show than go—on MR2s anyway, they simply are not necessary. But they are about as cool as spark plug wires can get. Photo courtesy Brian Hill.

The boost levels produced by your forced-induction SW20 may be controlled by either a manual or electronic boost controller. Each have their pros and cons.

Too cold, and you're moving backwards.

Typically 3S-GTE tuning utilizes spark plugs in the range of 6–9, depending on your specific ignition needs. The factory recommended BKR6EP-8 have a heat range of 6 (the last number indicates the gap as a function of 0.4, so an 8 would be a 0.32" gap, while an 11 would be 0.44").

Note: As a general rule of thumb, higher boost levels require smaller plug gaps. Gapping new platinum plugs is fine—and even sometimes necessary, depending on the plug—but do not gap used platinum plugs, or pre-gapped plugs.

Boost Controllers

The Idea: Increasing boost levels on a turbocharged car

Caution: Increasing the boost on any forced-induction car requires careful attention to a number of factors, namely general engine and clutch/transmission maintenance, tuning in regards to timing and air/fuel ratios and exhaust gas temperatures, and gasoline octane levels. In the staged upgrade chapters, recommendations are made with these considerations in mind.

Also, as a fail-safe, turbocharged cars usually feature a fuel-cut to protect the engine from malfunctioning turbocharger system components. Before increasing boost levels, these fail-safes must be disabled.

Boost Controller Basics—Boost is controlled on a turbocharged car by the wastegate. For now, understand that increasing boost levels is achieved by manipulating when this wastegate opens and closes. The longer it stays closed, the more boost that is produced.

On the SW20, boost is controlled by a

The 3S-GTE uses a wastegate as the final say in determining the amount of boost pressure produced by the CT26 turbocharger. Increasing these boost levels basically boils down to tricking the wastegate into staying shut longer.

partnership between the T-VSV, the ECU, and the turbocharger wastegate. T-VSV stands for turbo vacuum switching valve, not to be confused with other VSVs—vacuum switching valves—elsewhere on the car.

The factory SW20 ECU sets "low" boost pressure to 7–8 psi. Whenever ideal circumstances exist, such as a lack of detonation and appropriate relevant temperatures, the ECU charges the T-VSV and "high" boost mode is allowed, enabling the 3S-GTE will produce 10–11 psi. If you want to increase the boost levels beyond these meager amounts, a boost controller is necessary.

Manual vs. Electronic Boost Controllers—Boost controllers come in two kinds: manual and electronic. As you'd probably guess, the manual boost controller (MBC) is simpler and less expensive. It comes in two basic types: bleed valve

Toyota MR2 Performance

BOOST GAUGES

The first gauge you need in any forced-induction application is a boost gauge.

Beyond a graduated measurement of boost pressure, a boost gauge is also an excellent instrument of diagnosis, telling you exactly what boost you're making, by what rpm. The 3S-GTE has many built-in self-preservation features that it enacts if it detects non-ideal conditions in regards to intake temperatures, detonation, and boost, utilizing the fuel system, the TVIS system, and ignition timing all to help keep your engine together. A boost gauge is like an extension of your butt-dyno: once you've driven your MR2 Turbo enough, you can probably feel an unhappy ECU right away. However, running a boost gauge allows you to confirm what your butt is telling you, and clears the way of safe, precise monitoring of boost levels as you proceed with your buildup.

Peak-Hold Function—A useful option many gauges are available with is called "peak hold." Peak-hold gauges remember the peak reading recorded during a session, a third gear pull for example. At higher boost levels, third gear comes and goes pretty fast, and staring at a boost gauge when your MR2 is screaming down the road isn't the easiest or smartest thing to do. With a peak-hold gauge, you can keep your eyes out over the hood and on the road until you've pulled over to check your newfound data.

Many gauges also are available with features like warnings and playback as well. Playback is superior even to the peak-hold devices. Instead of simply recording the peak value reached, it will play back the entire sweep of the gauge needle, so you can tell if that 18 psi it recorded was a temporary spike, or was sustained. The APEXi 403-A045 EL Electronic Boost Gauge, for example, features 30-second playbacks—very useful.

Note: Exhaust gas temperature (EGT) gauges are helpful in roughly determining the air/fuel ratio you are running. Long-term, precise monitoring of your air/fuel ratio should ultimately be done with a proper wideband O_2 sensor.

The factory (OEM) blow-off valve on this Gen III 3S-GTE is visible at the top of the image, just to the left of where the strut tower bars intersect. There is no reason to replace your factory blow-off valve unless it is malfunctioning—in which case, replace with another factory valve. Photo courtesy Jensen Lum.

and ball-and-spring. Bleed valve types are incredibly simple—a pneumatic device that operates on air pressure principles. By using a bleed valve to reroute small amounts of air, wastegates can be fooled and boost levels can be increased. Ball-and-spring manual boost controllers improve on the bleed-valve principle by allowing the wastegate to remain shut longer. Why is this important?

Assume that an internal wastegate is designed to provide 12 psi of boost pressure. In cases like these, the wastegate itself may begin opening as soon as 6–8 psi, reducing the spool of the turbocharger. A ball-and-spring manual boost controller doesn't suffer from this issue, so that the control of the air is not a gradual and wasteful process, and boost is allowed to build more quickly. A ball-and-spring boost controller offers excellent bang-for-the-buck, and is a fine choice if you don't mind getting out of your MR2 and popping the engine lid each time you want to manipulate boost levels. The fact is, cockpit control of boost pressure comes in handy more often than you'd think.

The pros of the manual boost controller compared to electronic boost controllers are the price and simplicity. Simple is almost always good. There are cons, though, namely the location of the manual boost controller in the engine bay, where adjustments require under-engine-lid access. This, as well as the convenience of high and low settings, make electronic boost controllers excellent overall choices for the average turbocharged MR2 owner.

Note: For every 1 psi increase in boost, you should see about 10 whp, provided you are within the efficiency range of your turbocharger.

Blow-Off Valves

The Idea: Bleeding off trapped boost when the throttle closes

Blow-off valves, or bypass valves, are basically one-way valves that vent off boost pressure within the intake tract after the throttle is closed. This protects the turbocharger from compressor surge, where the air—basically stuck—is ingested by the turbocharger, rapidly decelerating the wheel and imposing physical stress on the unit itself. This is bad, so manufacturers, including Toyota, generally equip their turbocharged cars with such bypass valves.

The problem here is that Gen II 3S-GTEs—the engine found in the vast majority of all turbocharged MR2s—has its air metered by an airflow meter, or AFM. With the stock bypass valve, this is not a concern. The air is simply vented and plumbed right back into the intake tract, and there

BASIC MR2 PERFORMANCE MODIFICATIONS & THEORY

3S-GTE FUEL CUT

Even with a boost controller, the SW20 is not allowed by its ECU and associated cohorts (chiefly the T-VSV) to run beyond preset boost levels—12 psi for 1991–1992, 16 psi 1993–1995. There are a variety of options here to defeat this safety feature.

Some folks simply unplug the MAP sensor (located at the "rear" of the engine compartment at the wall separating the engine bay from the trunk, near the engine lid latch mechanism) and plug up the hose, but this option has significant drawbacks, namely the newfound ability to boost from here to infinity—not a good thing. It also keeps your factory boost gauge from working, if you're still using yours. For these reasons, this is not a wise choice.

Better options include a zener diode homemade circuit fuel-cut defense, or an adjustable device from GReddy or HKS, called the BCC and FCD respectively. Either way, before the boost can be turned up, some way to defeat the tattling electronics must be applied. For more information on the zener diode options, probably the least expensive solution for removing the fuel cut on your SW20, Google-search "zener diode + fuel cut." This method has been pioneered by other folks in the MR2 community, and supporting their work in that way is appropriate.

Note: Though Toyota saw fit to include MAP (manifold air pressure) sensors in the engine bay, that doesn't mean your USDM Gen II 3S-GTE is MAP-sensored. The MAP sensor is used to read boost, not airflow at the intake; that is the responsibility of the AFM (airflow meter).

Note: On the only other factory forced-induction MR2—the AW16—the ABV must be modified before increasing boost. See the staged tuning chapter for the AW11 for details.

This GReddy blow-off valve is an application-specific "bolt on" unit. There are also less-expensive, universal-style blow-off valves if you've got access to welding services. Photo courtesy Jensen Lum.

is never a temporarily rich condition. No harm done.

Aftermarket blow-off valves, however, are generally designed with the intention of producing a loud whooooosh sound, which is produced by venting the air back to the atmosphere. This air is basically unaccounted for; rather than being used in the combustion process, it effectively disappears, causing the engine to briefly run very rich. This hurts power, and might cause your exhaust to shoot a nice, big ball of fire. Good times.

3S-GTE Stock Blow-Off Valve—Blow-off valves vented to atmosphere (BOVs) can make your car sound loud, fast, and furious—whether or not this is a good thing is up to you. Unlike BOVs from older Mitsubishi product, for example, stock 3S-GTE Toyota BOVs works extremely well, and do not leak as boost pressure is increased. Therefore, aftermarket blow-off valves are almost never necessary, nor do they provide a performance increase over a properly functioning stock valve.

If you are running enormous boost pressures and you find your stock bypass valve is actually leaking, either repair or replacement is required. In this case, another stock valve, or even an aftermarket valve routed back into the intake tract is ideal. If you want to vent it to atmosphere, be sure that your tuning accounts for this lost air.

BASIC ENGINE MANAGEMENT PERFORMANCE
Tuning the Stock ECU

The Idea: Remapping the stock ECUs ROM for performance, and/or to control larger injectors

A ROM-tune (Read-Only Memory) is basically a reprogrammed stock ECU, and in most MR2 cases applies to the 3S-GTE engine. The 1ZZ-FE and 4A-GE, being less potent and normally aspirated, benefit less from this modification, and as such are less common in these engines.

Rather than replacing the entire ECU with a full standalone system, which would be an advanced upgrade, tuners—ATS Racing, specifically—can directly alter the maps written in your stock Toyota ECU. ATS can modify the maps on the factory ECU by use of a Techtom ROM board. ATS then

Toyota MR2 Performance

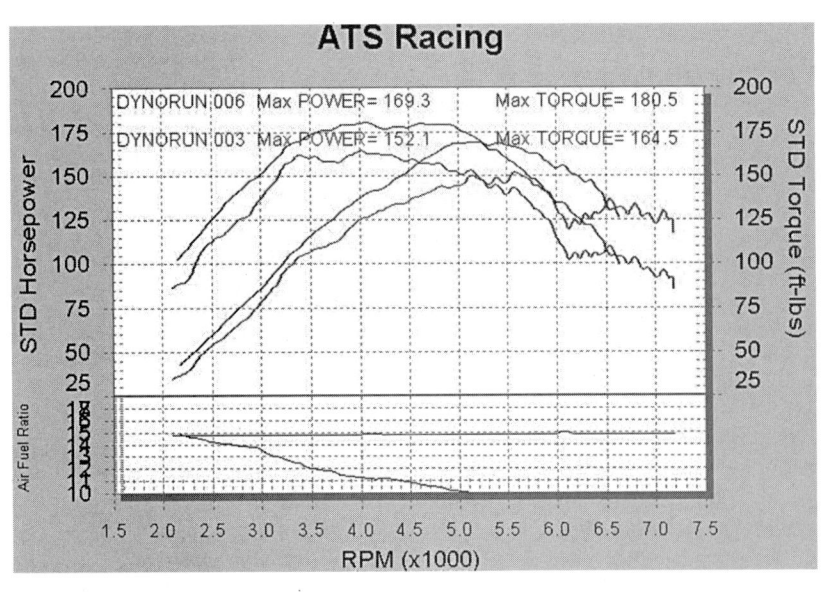

17 horsepower and 16 lbs-ft. of torque were gained by an ATS Racing ROM-tune—no rewiring, or complex programming. The ATS ROM-tune was a boon to the average SW20 owner seeking performance gains. Image courtesy ATS Racing.

The factory ECU contains air/fuel ratio and timing maps which can be optimized for additional horsepower. Unlike other cars whose ECU can be reflashed to increase boost levels, boost levels cannot be modified via ECU modifications on the MR2.

writes an "off-the-shelf" tune to the board, or, if they have the car on site in Texas, can write a custom, dyno-proven map on the spot.

Benefits of ATS Racing ECU ROM-tune
- defeat fuel cut, enabling higher boost levels
- raise rev limit
- enhance fuel and timing maps
- enable control of 550cc injectors
- dual-map options for lower and higher-octane fuels, valet mode, etc.

Note: Should you make future modifications that affect the volumetric efficiency of your engine after getting a ROM-tune, you can have it re-tuned again for a fee usually significantly cheaper than the cost of the original tune.

A ROM-tune is an ideal choice for the MR2 owner either on a budget, or not interested in the complex world of standalone engine management systems. If you're shooting for the popular 300 whp level, it is hard to fault a set of 550cc injectors and a ROM-tune by ATS. ATS Racing in Texas has been the most serviceable source of such tunes for MR2s for years.

BASIC INTAKE PERFORMANCE
Air Intake Modification

The Idea: Removing air intake restrictions, especially at elevated boost levels

The air intake system is composed of the factory airbox—which houses the air filter, airflow meter, and the air inlet pipe that connects to the turbocharger. Typically, when seeking to gain performance, usually the stock air filter and air filter housing is the first to go, replaced by a cone filter of some kind. As with exhaust modifications, most intake modifications apply to the 3S-GTE (SW20).

Airflow Meter—The airflow meter, or AFM, has been a surprisingly important player in MR2 tuning for years, not so much on the AW11 as on the 3S-GTE-powered SW20. This relatively simple device is basically a housing used to measure the amount of air being ingested by the engine. To this value the correct amount of fuel can then be applied to produce an appropriate (for load and overall driving conditions) air/fuel mixture.

The door-flapper valve that moves in response to airflow must look very restrictive, because for years the aftermarket has been trying to get rid of it, with products from companies as accomplished as HKS—the HKS Vein Pressure Converter, or VPC, often used on SW20 setups back in the 1990s—developed to replace it. Eventually it became clear, though, that the AFM wasn't a huge restriction, with several MR2s running deep into the 11-second range on the factory AFM setup. Now certainly the amount of air that needs to be ingested by an 11-second (350+ whp) engine probably does surpass that amount for which Toyota engineers designed the stock AFM—and there becomes a time when air metering must be enhanced; just because a car can make 400 horsepower with part X doesn't mean it wouldn't be

LITTLE BLACK BOX: APEXI S-AFC

MR2 Turbos, like most production cars, run very rich as setup stock because a rich fuel mixture, up to a point, is generally safer than one that is lean. Rich fuel mixtures also hurt horsepower, and leaning out that rich condition—that is, removing fuel—can add some of that power back. This problem of too much fuel is only made more immediate when you add larger injectors. The OEM ECU doesn't know the larger injectors are there; it assumes it is still controlling stock Gen II 3S-GTE 440cc injectors, and so some manner of controlling the ECU's relative ignorance must be had. There is simply too much fuel per pulse.

Before ATS Racing got around to amending OEM ECUs to control larger injectors, the easiest, most inexpensive method of controlling that extra fuel was an APEXi S-AFC. The S-AFC is what is called a "black box"—any piece of hardware, with imbedded software, that is intended to go over the ECUs head and alter its direction. The problem is the way the S-AFC and the MR2 interface.

The S-AFC basically lies to the ECU, altering the reported signal from the airflow meter (AFM) to the ECU. What signal it reports depends on whether you are attempting to lean or richen your mixture. Assuming you are attempting to control larger injectors, it tells the AFM that there is less air than there actually is, subsequently convincing the ECU, and then the injectors, to produce less fuel. In theory, it sounds great. The problem is how the ECU reacts to that reduced signal. Instead of easing back the timing, it advances the timing, and Gen II 3S-GTE ECUs already possess quirky stock timing maps. This added timing can kill your engine fast.

This type of problem is why piggybacking black boxes are gradually being replaced by more fluent standalone engine management systems. Simply altering signals and lying to systems components to render an end was never an ideal way of doing things, especially considering the crucial nature of engine management. Of course the standalone has issues of its own, namely complexity of install, tuning, and its entry price, but the black boxes never completely filled the need for piggyback MR2 tuning. In the free market, if there is a need, there are generally product designers and retailers looking to fill that need.

There is a relatively safe way to use the S-AFC. Rather than simply wiring one up to do all the work alone, it should be used in tandem with an adjustable Fuel-Pressure Regulator (FPR), and a fuel-pressure gauge to let you know what's happening. An adjustable FPR allows you to manipulate the fuel system's pressure. The fuel is delivered to the rail with a certain pressure, which is dictated by the pump, the FPR, and involves the bore of the fuel rail and other criteria as well. This is called base fuel pressure, and turning it down effectively reduces how much fuel is being introduced into the cylinders.

So if it's that easy, why not simply install bigger injectors and turn down the base pressure? Appropriate fuel pressure is required for the fuel to properly atomize in the combustion chamber—image a hose issuing water freely, and a hose forcing water out with your thumb somewhat covering the end. The former—that relative dribble of water—is what an injector would look like without adequate pressure. Injector design is a very precise science, and engineers spend a lot of time optimizing spray patterns, pressures, directions, and so on through nozzle and injector development. Ruining this design with crude adjustments can rob you of some of the horsepower you were seeking in the first place. To use the S-AFC in an MR2 Turbo with larger injectors, you must first adjust base pressure to 30 psi with the FPR, and then use the S-AFC to remove fuel in small increments, from a +10% range to a 5-8% range at appropriate rpms, depending on your exact setup. The important thing is to never go less than stock—0%—which will keep the AFM and the ECU from butting heads and blowing up your ringlands.

making 425 with part Y. Still, the AFM has been vilified in many tuning circles as a major restriction in almost immediate need of replacing, and that simply isn't true for the majority of light-to-moderate-tune 3S-GTEs.

Mazda Miata folks have postponed a similar issue by using certain Mazda RX-7 AFMs to more responsibly meter larger amounts of air, but these upgrades only work with other alterations, and most importantly an original need for them to begin with. Simply installing a set of larger fuel injectors will make your MR2 run so rich (read: too much fuel) it may not, in fact, run at all. The need for, and control of, more fuel must exist for extra horsepower to indeed be made, and the same is true with air. Throwing on a bigger AFM isn't going to magically make more horsepower. Always seek the safest air/fuel mixture that renders the most ideal power curve for your application—and seeking that mixture implies considering the

This 3S-GTE AFM has an APEXi air filter installed on it, mostly hidden from view by the heat shield.

4A-GE INTAKE MODIFICATIONS

As we address in Chapter 5, 4A-GE (AW15) intake modifications are fairly low in priority. There simply is not much gain, and the money is better spent on track time even if you've already performed other, higher-priority modifications like sway bars, urethane bushings, and sticky tires.

However, at later stages of modification, when you're trying to extract everything it makes sense to extract, looking for 5 hp with an aftermarket intake isn't unreasonable. The basic intake here is an AFM adapter that allows you to bolt a K&N or other quality performance air filter directly to the AFM. Obviously, moving away from the engineering in the factory airbox forces you to lose something. In this case you are sacrificing quality of air filtering, and increasing the potential of catastrophic engine failure form ingesting water. These are authentic concerns that require other adjustments, especially considering where you live and drive your MR2: rain and water isn't a primary concern for those in southern California, but grit and salt from the beach and ocean can be.

The main power potential in 4A-GE intakes at stock or near-stock power levels lies in the length of the intake tube itself. Just as the revised 1987–1989 AW15 routed the intake through a convoluted path behind the trunk and even into the fenderwell, presumably in the name of a specific harmonic resonance, you too can roughly make similar modifications. Depending on other engine modifications—camshafts or exhaust size, for example—you can alter the length of the intake in order to create a certain resonance frequency that does, in fact, impact the powerband.

Many OEMs, including Toyota, experiment with artificially lengthening and shortening the intake path within the intake manifold itself for a similar result. If your MR2 is designed for performance at a certain rpm, experimenting with the length of your intake tube to complement your setup can be worth the effort. I've seen up to 5 hp and 8 lbs-ft. of torque from this sort of experimentation, though rarely across the entire powerband; remember, whatever the resonance that increased torque, may have a detrimental effect on upper rpm performance. What you're effectively doing is choosing a portion of the powerband you want to enhance, sometimes forgoing performance elsewhere in the process.

Though the intake noise can be pleasing to some, considering the potential for lost power, lost filtering capability, and lost money from the modification itself, for most 4A-GE owners, a stock airbox is your best bet.

Note: If you can find a drop-in K&N filter, at bare minimum you'll have a reuseable air filter that you won't have to replace. You also might see 1–2 hp on a dyno, though most dyno-testing I've seen on this modification has been inconclusive.

All that is necessary for a basic air intake modification is an air filter, and an appropriate adapter to connect the filter to the AFM. This is the APEXi unit shown.

tandem of both fuel and air together, which drags engine management into the issue. Some method of controlling and manipulating these signals—the factory ECU, or some arrangement of aftermarket devices must be sought out, and that is where your attention should be focused.

Throttle Body Inlets—The 1991–1992 SW20 had a tapered throttle body inlet that reduced around 3-5 whp, the cone shape of this early inlet placing a bottleneck right at the throttle body. Without changing the factory horsepower rating of the 3S-GTE, Toyota themselves changed to a non-tapered, straight-through design in 1993. The straight-through design of the later inlet removes this airflow bottleneck, especially important in turbocharged cars running boost levels beyond stock.

ATS Racing has been selling these inlets for years, producing a handsome piece of billet aluminum. You can also simply bolt on an stock inlet from a 1993–1995 SW20, but considering the low cost of the ATS Racing piece, simply ordering one from them is ideal. This modification only applies to the early 1991–1992 SW20.

BASIC COOLING SYSTEM PERFORMANCE

The factory cooling system on any car is a complex mess of sensors, hoses, heat exchangers and fittings. On a mid-engine car it is especially so, and when you turbocharge that mid-engine car (as in on the SW20), so much more heat is generated that the cooling system must not only be stout, but consistently monitored through methods beyond simply watching the coolant temperature gauge. Two easy ways are to visually inspect all coolant systems regularly—with oil changes if at no other

Basic MR2 Performance Modifications & Theory

The door-flapper is visible in this photo, curiously a source of consternation for many SW20 owners. It certainly can become a restriction at elevated power levels on the 3S-GTE, but until you've gotten to those levels, leave it alone.

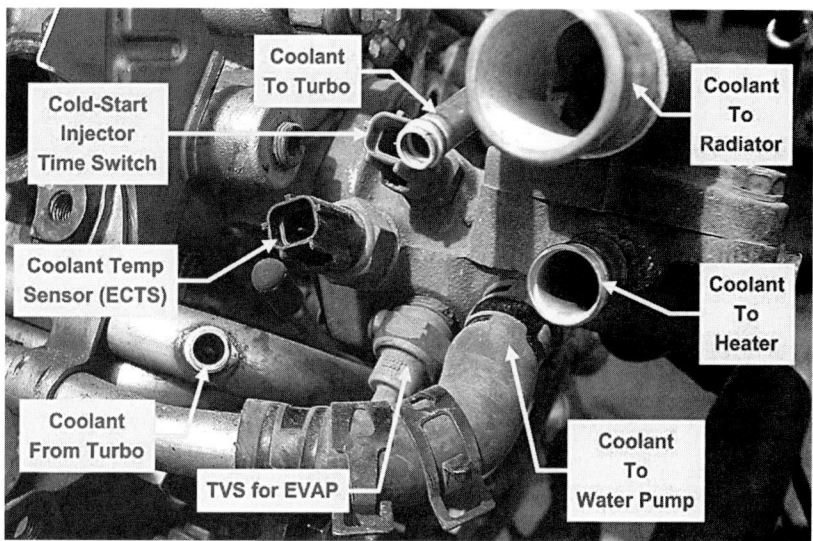

Like all MR2 cooling systems, the cooling system on the 3S-GTE must carry the coolant from the engine all the way to the front of the car, where the radiator is, and then back again. This makes understanding the 3S-GTE cooling system crucial on modified SW20s.

time, and always be alert for that unmistakable, sweet smell of leaking coolant.

Like most production cars, every generation of MR2 features a stock cooling system with a significant amount of margin built in, so upgrades to radiators and other cooling system components are generally unnecessary. As we've discussed, a low-temperature thermostat is only as capable as the cooling system it is serving as a gateway for, and even then, running engines indiscriminately colder than stock thermostats are designed for, just because a "colder thermostat" is available for your car, is not smart. Colder isn't always better.

As a general rule, modifications to the cooling system are rare on MR2s. Like the rest of the car, the cooling system is extremely robust, and the majority of your energy should be spent here maintaining the stock Toyota setup in OEM healthy condition.

Thermostats

The Idea: When necessary, lowering the coolant temperature

The thermostat is used to maintain the coolant temperature within a certain range by controlling the flow of liquid. The rating is the temperature that the thermostat will open, increasing the coolant flow as the engine and coolant heat up. Too much heat generally has a negative impact on performance. But how much is "too much" depends.

Low-temperature thermostats are popular in the domestic performance aftermarket. For example, a "160°F T-stat" is a thermostat designed to open at a temperature less than the factory thermostat, in theory allowing the engine to run cooler. It'd seem like a cooler-running engine would be a good thing. It's not that simple, though.

First off, the cooling system has to have the capacity to meet that thermostat's suggestion. The only thing the thermostat decides is when to open or close—it can be thought of as a gate to the cooling system. Big horsepower makes big heat. No matter what temperature thermostat you install, you're only going to run as cool as the capacity of the cooling system.

Second, engines are designed to operate most efficiently at certain temperatures, generally around the stock thermostat's setting. Obviously, running too hot yields overheating, universally accepted as bad for performance. But to a point, an increase in the engine operating temperature can improve performance by allowing ignition advance—if you tune for it. Running too cool can be bad, too. Emissions, wear, efficiency and power output all suffer considerably if the engine doesn't reach proper operating temperature. In general, a properly functioning Genuine Toyota thermostat is the way to go.

Note: An old-school shade-tree racer's trick was to run no thermostat at all under certain conditions. In effect, this is akin to a thermostat running open all the time, causing the engine to take considerably longer to properly warm to operating temperature. While this is an improvement over a stuck-closed thermostat—possibly allowing you a drive home if a stuck thermostat had been causing you to

Toyota MR2 Performance

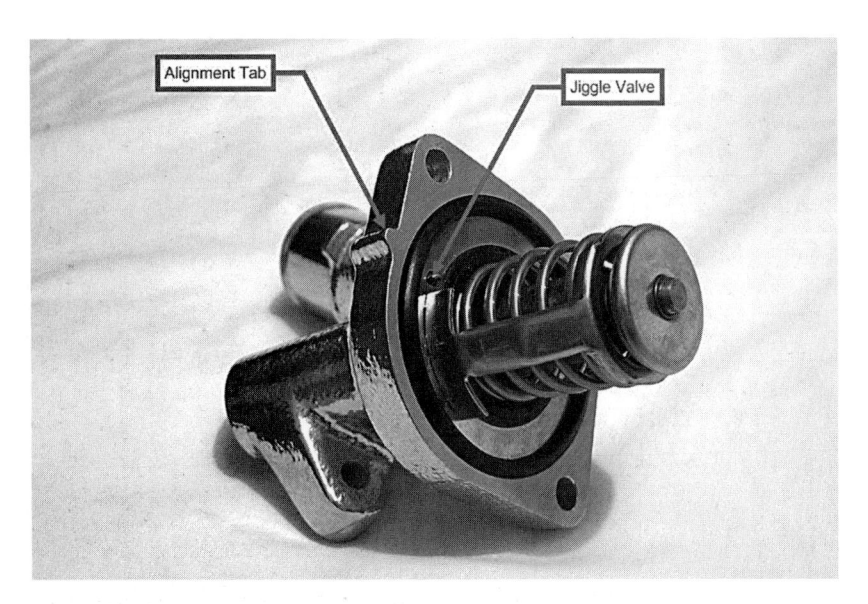

Before you choose to run a colder-than-stock thermostat, be sure you've done your homework. Indiscriminately choosing a colder thermostat because some aftermarket company sells one for your car isn't smart. When you do need to install a new thermostat, be sure the jiggle valve is aligned with the alignment tab on the thermostat housing.

overheat, stranding you on the side of the road—this is generally not recommended for a street-driven vehicle.

High-Pressure Radiator Caps

The Idea: Raising the pressure of the cooling system raises the coolant's boiling point

There are ways to enhance your cooling system's ability to do work when it becomes necessary, e.g. you've doubled the horsepower the factory components were designed to serve. One is a high-pressure radiator cap.

Radiator caps, even factory caps, have a pressure they impose on the cooling system. This pressure provided by the stock cap increases the boiling point of the water by 40–50°F over what it would have without that pressure. A higher pressure, to a point, will increase that value further.

TRD and TOM's both manufacture higher-pressure radiator caps for most MR2s. The problem, though, is that all that extra pressure also places increases wear on cooling system components, which can become a serious issue on cars with a lot of miles. Pressure in the cooling system means pressure everywhere. This isn't usually a significant worry on most applications, but it can be. Whether or not you choose to include a high-pressure radiator cap in your build or not is a personal choice really based on how much you trust your engine and cooling system. If they're in good shape, there is no reason not to run a high-pressure radiator cap. An alternative to increasing coolant boiling-resistance with a high-pressure cap is waterless coolant.

Evans NPG Coolant

The Idea: Using waterless coolant, which is less likely to boil

When coolant boils, it creates air pockets in the cylinder head, which allows for those areas to get hotter than those in contact with coolant. While such boiling allows for tremendous amounts of energy—and thus heat—to be released, hot spots are extremely undesirable. Hot spots create trapped vapor, which causes detonation, and we know what detonation can do. Several Gen II 3S-GTEs torn down in recent years have had evidence of boiling in the cylinder head, and would-be tuners set about to fix that. Some theorized that water pump cavitation was the problem—the stock water pump could not keep up with pumping demands at higher rpms. Others thought the water jacket design through the head was the issue, and indeed this was one of the changes Toyota made when they developed the Gen III 3S-GTE. One theorized fix was simply a coolant that would not boil as easily.

Evans Cooling Systems is a leading manufacturer of high-performance cooling system components and fluids for both racing and street use. They market a revolutionary line of coolants under the banner NPG, for non-aqueous proplyene glycol, that use no water at all. NPG+ and NPG-R. NPG+ has a boiling point of 375°F, much higher than a standard coolant/water mix. NPG+ requires a pressureless system, which means no high-pressure cap from TRD, and reduced stress on the cooling system components, which is a good thing.

NPG Install Notes—It must be noted that NPG+ is a major pain to install. Water, because it has a relatively low boiling point, will interfere with the NPG+ coolant's ability to resist overheating. The entire cooling system, which is considerable in any MR2, must be flushed completely—and when you think you've done so, you should flush it again. This means the heater core as well, and those hoses are tons of fun to remove, let me assure you. (A pair of 90-degree needle-nose pliers go a long way in helping remove them.) If you're using NPG+, which is already very pricey, this can add up fast. It's a good idea to flush your cooling system (including the heater core) with a less expensive waterless coolant like Sierra brand coolant until you've got all the water gone. You should also send Evans a sample of your coolant to be sure you've done it all properly when you've finished.

Basic MR2 Performance Modifications & Theory

Catch-Cans

The Idea: Keeping things clean—especially your intercooler plumbing on turbocharged cars

Hot gases from swirling internal engine parts and piston blowby move about your bottom end, looking for somewhere to go. With a catch-can, crankcase gases are caught by can, where the oil from the gases condense and collect, rather than the route Toyota established, which soaks your intake pipes and with a thick coat of grimy, intercooler efficiency-diminishing sludge. A simple filter, from K&N for example, can be used to vent these gases as well, but the GReddy catch-can works well and looks good.

For the Gen III version of the 3S-GTE, Toyota installed a factory catch-can, the innocuous looking black rectangular object seen in this photo. When the factory makes modifications, it is always worth strongly considering following suit. Too bad the United States never saw the Gen III in a production SW20.

Chapter 4
Advanced MR2 Performance Modifications & Theory

Being mid-engine, rear-drive, and relatively lightweight, MR2s of any generation make excellent high-performance platforms. To save yourself a lot of time and money, though, you must first have a grasp of what does what. Photo courtesy Hsun Chen.

Now that we've laid the foundation for creating a basic balanced performance car in Chapter 3, in this chapter we'll take a look at more aggressive parts that produce even higher levels of performance. These modifications are simply a progression on the fundamental ideas of performance theory laid out in the previous chapter. Some of these parts would replace the more basic modifications, coil-overs replacing shock-and-spring combinations for example, while others are entirely new, such as engine management systems. So what makes the parts below more advanced? In most cases, it's an issue of either cost or compromise.

Almost everything—in cars, and in life—is a compromise; you give a little, you get a little. Major manufacturers like Toyota are masters of that compromise, so if it's middle-of-the-road compromise you want, stock is finished. But you may look at compromise as slack in the chain. If so, that compromise is where OEMs leave room for performance. Most of the parts discussed here in Chapter 4 require either fairly complex installation (urethane suspension bushings), "programming" (aftermarket camshafts need to be degreed), require significant changes elsewhere in the car (big turbocharger without an internal wastegate needs one), or have higher costs associated with their purchase and use (engine management systems or engine swaps). Still, none should be beyond your consideration or imagination. Even if your MR2 is a dead-stock commuter and always will be, read on for morbid curiosity if nothing else. Forcing yourself to look at and consider options you closed your mind to previously is a talent unfortunately many do not have.

Reviewing the common pathways to increased MR2 performance can only expand your current convictions, so let's take a look at those parts that can make your MR2's performance even mo

INTERMEDIATE & ADVANCED HANDLING PERFORMANCE
Coil-Overs
The Idea: Shock-and-spring assemblies with matched rates that are ride-height adjustable

Coil-overs are full "corners" of a suspension—a shock body accompanied by a wound spring that is ride-height adjustable through an adjusting threaded collar. Coil-overs have several advantages over traditional shock/strut and spring setups, including lighter weight and liberal ride-height adjustment. Some models of coil-over even offer diverse alignment (camber) adjustment via special mounts. But the most significant performance gain offered by coil-overs lies in the ability to properly corner-weight your MR2.

In most printed automotive media, discussion of weight balance is limited to ratio of physical weight front to rear, the holy grail being a "50/50 weight balance." This is certainly important, with a similar balance helping make your MR2 feel different than most other cars on the road. Coil-overs are relevant to a different kind of balance, though.

Coil-overs, being independently adjustable by that threaded collar at each corner, also allow lateral and crossweight adjustments to be made, so we're dealing not with a physical weight balance front to rear, but a "corner"

ADVANCED MR2 PERFORMANCE MODIFICATIONS & THEORY

Advanced planning can help promote advanced performance on the street and track, even for generations of MR2 out of production for almost twenty years now. Photo courtesy Hsun Chen.

Coil-overs replace both the shock/strut, and spring with adjustable parts with matched rates. The threaded collar and spring "perch" are visible in on these JIC coil-overs. Photo courtesy Mike Choi.

balance. These adjustments affect how much weight each corner exerts on the road. Manipulating these measurements—front to rear as well as left to right—can impact the handling balance of your MR2 greatly.

Why is this important? Simply put, just as many owners will move batteries to the front or rear of the cars to help balance front to rear weight bias, cars need to be balanced laterally as well. A car lacking optimal lateral balance might be significantly slower in right turns than left (or vice versa), which, of course, is not ideal. In the interest of keeping things predictable and easy to drive fast, there is a need for balance equity. Like some of the more complex aspects of performance, engine management tuning for example, proper corner and cross-weighting is best left to either a reputable suspension shop, or by your own thoroughly researched and educated devices. As easy as it may be to improve these measurements—your "corner weights"—it is far easier to make it worse (more on this later). Either way though, it is important that you at least have a fundamental understanding of what it is involved in this facet of the black art of suspension tuning.

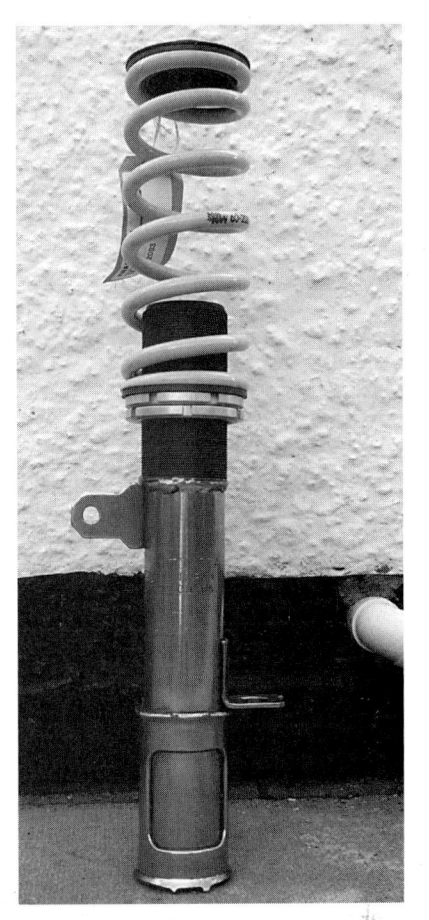

The Corner-Weighting Process—In a nutshell, corner-weighting is the measure of the amount of (static) weight on each wheel. After calculating total vehicle weight, these individual "corner" weights are then expressed as percentages of total vehicle weight.

We know that coil-overs are ride-height adjustable, this adjustment coming through the turns of the adjustable spring perch. When you alter the height, it has a direct impact on the weight that wheel places on the road. Ideally, the weight on the front left wheel will be equal to the weight on the right rear, implying an equal weight front to back, and side to side.

Once taken, cross and rear weight measurements can then manipulated by adjusting coil-over ride-height at separate corners, and by moving ballast—mobile (though securable by some means) physical weight—around in the car to obtain desired results. In a nutshell, this is how coil-overs enable corner-weighting to be done.

So Who Needs Coil-Overs?—So coil-overs are wonderful in their adjustability and potential for accurate matching of the spring and damper rate. They can also allow for ride-height adjustment to meet many applications—low for show, higher for daily driving, and somewhere in the middle for high-performance work. What's the catch? Essentially, what makes them great—their adjustability—is also what can make them a deal-breaker for some folks, too.

Coil-overs are considerably more expensive and complicated than the basic performance suspension setup accessible to most owners, i.e., an aggressive alignment, sticky tires, adjustable shocks and

Ride height is adjustable on coil-overs via a threaded collar. Custom spring rates can also be chosen to fit your need. Photo courtesy David Jones.

59

TOYOTA MR2 PERFORMANCE

Coil-overs not only enable precise matching of spring and damper rate, but also allow for ride-height adjustment at each corner, eliminating awkward rakes, as the beautiful stance on this SW20 demonstrates. Photo courtesy Jensen Lum.

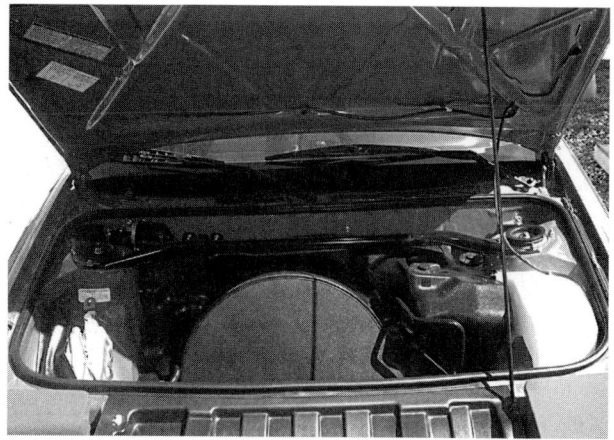

Though the MR2 is mid-engine and rear-drive, it's still the front wheels that do the turning, and so the power-steering components are still placed in the front-trunk. On the AW11, no power-steering means no power-steering components—just a spare tire. Photo courtesy Helder Carreiro.

COIL-OVERS AND RIDE QUALITY

A set of quality coil-overs with well matched spring-and-damper rates should not significantly impact ride quality—in fact, if you're replacing a setup that was overly stiff and poorly matched, ride quality should improve with a set of coil-overs. While what you consider an "improvement" is subjective, understand that it is not a a foregone conclusion that coil-overs will automatically ruin the ride in your MR2.

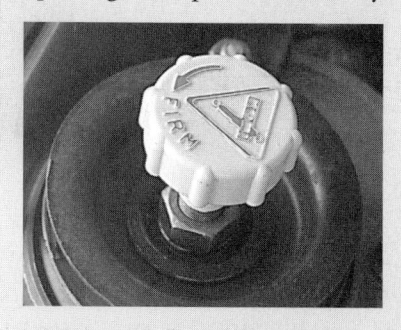

This of course depends on whether you use a solid pillow ball mount or not. A pillow ball mount is an aftermarket style of suspension mount designed to replace the factory suspension mount. As is the case elsewhere throughout the suspension, the factory piece is soft, and the aftermarket piece is harder—in this case, the OEM piece Toyota used was a relatively compliant rubber. In the interest of handling and alignment precision through minimal deflection under load and accurate alignment settings, many aftermarket manufacturers offer pieces made of metal—and yes, this results in a ride has harsh as you'd expect. Though these mounts offer significant amounts of alignment adjustability, for the street-driven MR2 they represent too harsh an impact on overall ride quality.

Note: TRD of Japan used to manufacture a piece made still of rubber, but harder rubber, a middle ground between soft factory rubber, and rough, aftermarket pillow ball mounts. These pieces, when they can be located, are in high demand for obvious reasons.

The material beneath the shock adjustment knob is made of rubber on this MR2—and most likely yours, too. Replacing it with something harder—aluminum, for example—can improve handling and overall suspension precision, at the cost of ride quality. Photo Bryan Heitkotter.

adjustable sway bars.

Stock class autocrossers—allowed only R-compound tires, aftermarket shocks, an aggressive alignment and a front sway bar by most rulebooks—still manage to greatly improve the handling performance of their MR2s. Though coil-overs can indeed offer higher levels of performance because of the aforementioned tuning adjustability, such suspension tuning requires extensive time via trial-and-error. Any such potential goes in both directions: as easily as you can improve things, you can make them worse as well. Constantly making adjustments, and measuring gains through some means other than what "feels good" is crucial.

An MR2 owner considering coil-overs must also be willing to come to terms with spending perhaps thousands of dollars on a modification that will mainly benefit your MR2 in 9/10s driving conditions and beyond, i.e., the race track, and almost nowhere else. This is in contrast to performance tires, adjustable shocks, sway bars, brake pads, boost increases, etc., whose performance can affect a broader, more frequent range of performance.

Quicker Steering Rack

The Idea: Better feedback, quicker steering, lighter weight

The AW11 never had power steering, which is a good thing because it didn't need it. The EHPS (Electro Hydraulic Power Steering) in the SW2x was an advanced system featuring an independent electric pump, and thus no strain (read: power-robbing) on the engine. It also shuts off at higher speeds and was considered a brilliant system for

STEERING SYSTEMS AND WEIGHT

As mentioned, the AW11 never had power steering. Frankly, it didn't need it, so that is a good thing. And so for AW11 owners there are no weight or steering-rack swap considerations.

For the SW20, there were indeed power steering-related weight issues. Jeff Fazio, a very active MR2 enthusiast, and owner of "Hyde," the highly tuned SW20 referenced elsewhere in the book, weighed his EHPS components when he swapped them out for a non-power system and found that the EHPS system, including lines, motor and hardware—weighed in at exactly 32 lbs. and 3.0 oz, while the manual rack and all of its associated goodies weighed 13 lbs. and 4.1 oz. A little math says that the conversion saves just under 19 lbs.—not bad, but you've got to be desperate for a lightweight car for weight savings to be your primary reason to swap out steering racks, as the process, while not impossible by any means, is not an afternoon job.

The slower steering ratio of the non-EHPS rack in the SW2x is also a strong reason to stick with EHPS should your MR2 have it. For most owners, increased feedback and "feel" are the primary causes of such a swap, but again, this is subjective. Personally, I prefer the feel of the EHPS system over the non-power setup, but classic sports car theory tells you that for feel, feedback, and just-right sorting out, nothing beats a non-power assist steering system.

ZZW30 Steering—While it's possible for ZZW30 owners to consider ditching their EHPS in the name of weight-savings and increased feel as well, these topless MR2s are newer and generally worth more than previous generations. Basic economics what they are, there is more financial vulnerability making significant modifications to a more valuable car, so as is the case with much of the ZZW30 aftermarket, there are fewer options here. Since fewer folks have torn out racks and tried installing units from other cars to see what fits, what issues there are, etc., you're pretty much on your own.

Seeing as the ZZW30 rack is plenty quick at 2.7 turns lock-to-lock, the need is not great, either. In fact, it's a bit of a mystery why Toyota ever installed power steering on a 2200-lb. sports car so thoroughly designed for lightness that it lacks even a basic trunk.

The one thing that affects everything your MR2 does—accelerating, braking and handling performance—is weight. In pursuit of that lighter weight, many SW2x owners swap out steering racks to gain every tenth of a second on the track. Photo courtesy Hsun Chen.

such a relatively inexpensive car from a major manufacturer. For many folks, though, nothing beats a non power-assisted steering system for the increased control, steering feel and feedback. With some effort, steering racks can be switched out, so if you have power-steering and don't want it, there are options.

The relative speed of steering racks are usually stated in two ways: degrees of relative movement, or turns lock-to-lock. The SW2x EHPS rack-and-pinion features a fairly quick 3.2 turns lock-to-lock rack. Given in degrees, it works out to be 18.2:1; this steering ratio tells you that the steering wheel must be turned 18.2 degrees to move the front wheels 1 degree. The non-power rack features a slower 20.5:1 steering ratio.

Quicker Steering—As the name would suggest, a quicker ratio requires fewer turns from lock to lock—quicker steering. The factory rack in the AW11 requires 3.25 turns from lock-to-lock, while the rack from Quaife only 2.5. Although the car will respond quicker to steering wheel input, it may also require greater steering effort as well, though due to the mid-engine layout of the MR2, this is less of a problem than front-engine cars—there's simply much less weight over the steer tires.

For the AW11, Quaife manufactures a quicker rack-and-pinion ratio, and as far as I know is the only company to do so (retail $279, Quaife America part # 94.435.001, Quaife UK part # QSF22E003). The MR2 can thank its cult-classic cousin, the AE86 Hachi Roku for that. (The AW11 pinion is actually about 1/2" longer than the AE86 pinion, but there is enough flexibility in the tangent pieces—couplers, joints, etc.—that most owners have not had issues installing this piece from Quaife. Obviously, gains in subjective feel and objective performance make issues like these relatively inconsequential, provided they are indeed a simple fix.

Most aftermarket pieces are either larger, or firmer. This urethane suspension bushing is simply a firmer version of the stock piece, reducing compliance, and increasing feedback and general handling precision. Photo courtesy Mike Choi.

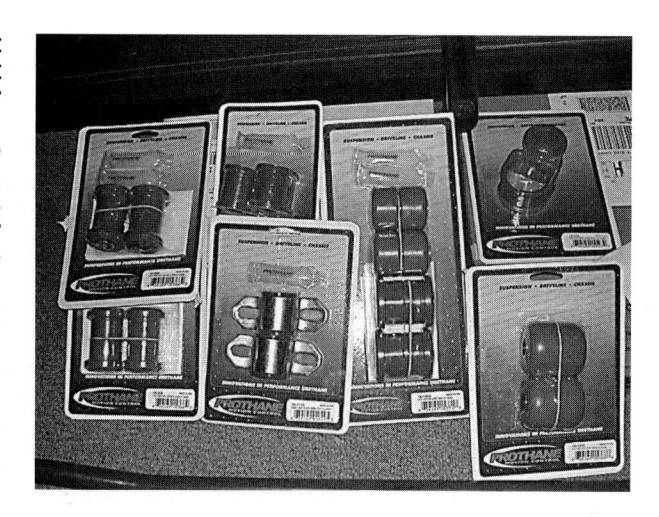

Suspension Bushings

The Idea: To further reduce deflection of the suspension.

Improve Steering Feel—Bushings are simple little components that reduce deflection of the suspension during cornering maneuvers, and due to their reduced compliance over softer (worn) rubber bushings, offer improved feel. They also require no special service once installed beyond, in some cases, periodic lubrication, nor or are they relatively expensive. So why are they in the advanced chapter? Because they are difficult to install.

Most suspension pieces are made of metal, their job to absorb road irregularities while enabling the tires to maintain optimal physical contact with the road through a variety of loaded and unloaded conditions. Stock bushings are typically made of a softer rubber as a compromise between handling and noise and ride harshness. But for performance cornering, you want as little movement or "deflection" of the suspension components as possible. Aftermarket bushings are made of either a hard rubber compound or polyurethane, while racing bushings are usually bronze or some other metal. These stiffer materials allow free movement of the various control arms, tie rods, end-links and anti-sway bars and absorb impacts. However, because they are stiffer, aftermarket bushings allow for more precise suspension geometry to be maintained throughout the suspension's travel.

Installation—Bushings are generally replaced along with more advanced suspension modifications because the installation is more difficult. Typical AW11 bushings suffer from more than two decades of road grime and grease. A lot of patience is necessary to replace them, so it's a modification that should be considered with a major suspension upgrade: your MR2 teetering on jack stands, suspension in an uncomfortable number of pieces, blowtorches and enough acrid smoke to get the EPA's attention. Still, to the discerning driver, the compliance and general slop allowed by rubber pieces of said vintage is unacceptable, and while the improvement you'll enjoy depends on the condition of the pieces you're replacing, any self-respecting car purported to "handle" well should have fresh bushings everywhere it matters—especially the suspension.

Limited Slip Differentials (LSD)

The Idea: Less wheelspin, more grip for acceleration, and acceleration-while-turning

MR2s don't have much history with LSDs. With all MR2s being relatively modest horsepower-wise, and being exceptionally well-balanced and rear-drive from the factory, the need for a LSD is not as great as it is with a powerful front-drive car, for example. USDM AW11s did not feature an LSD, and the SW20 only rarely featured a basic viscous setup whose performance didn't exactly set the world on fire. In fact, it wasn't until the ZZW30 that Toyota ever installed a LSD of any merit. So, if in the name of performance you're looking to add what Toyota left out, what do you look for? Before you begin, you need an understanding of LSD design. The nomenclature can get confusing, and an understanding of what each LSD type does will help you understand what you're getting yourself into.

What's an LSD?—A limited slip differential is a handy invention, a veritable savior for front-drive cars, but certainly useful on rear and all-wheel drive cars as well.

A differential is basically a mechanical means of allowing drive-wheels to travel different speeds, as the vehicle turns, for example. Just like a sprinter on an oval track, the inside tire has a shorter path to follow than the outside tire. Mechanically, there is a need for some "compromise." A differential provides this compromise basically by distributing each drive-wheel with an equal amount of torque. This is fine in a straight line, but in a corner, envisioning the inside-sprinter-on-an-oval-track analogy above, this does not work as well. A variety of forces are working uniquely on the inside and outside tire—unique weight, unique torque, unique traveling distances, etc. A limited slip differential basically detects slip, and redistributes torque to the wheel slipping the least, where it can be most effectively applied.

The B13 and B15 Nissan Sentra SE-R and the Acura Integra Type R all enjoyed stellar front-drive handling in large part due to their factory-equipped limited slip differentials. LSDs are basically a huge pile of gears following set contours, with a large of

LSD TYPES

In addition to the unique marketing naming manufacturers use for the same differential types, and the mechanically unique LSD styles (more on that shortly), LSDs are again separated by the circumstances under which they are active: acceleration, braking or a combination of both.

If you've heard the term 1-way, 1.5-way and 2-way in discussion of LSDs, it refers to that LSD's ability to operate under unique loads.
- 1-way: Active under acceleration only
- 1.5-way: Active as a 2-way, but with less lock under deceleration
- 2-way: Active under both deceleration and acceleration

Note: 1.5-way units offer added stability under deceleration over 1-way units, and are generally also reserved for progressive track cars that require stability under braking.

Note: Two-way LSDs are very aggressive differentials that are typically reserved for track-only cars.

torque acting upon them. Some LSDs can be very aggressive, feeling very dramatic—a helical unit in a torquey front-drive application can literally yank the steering wheel right out of your hand.

Other LSDs are more transparent in their feel. The 1991–1994 (B13) Nissan Sentra SE-R, for example, used a viscous-coupling limited slip differential. A viscous-coupling LSD uses a thick gel-like substance stuck between two sets of plates, each connected to an output shaft. The greater the difference in the speed of the rotation of the plates—which happens when one wheel is turning faster than the other, usually indicating "slip"—the more the fluid is sheared by the plates. When this happens, the slower moving plate—connected to the wheel with more grip—receives more of the torque that was previously being sent to the spinning wheel. This all probably doesn't sound very efficient, and that's because it's not. A viscous LSD's locking rate is generally very low, rarely able to transfer an adequate amount of torque between wheels for it to be something you go out hunting in a performance catalogue for. While certainly better than an open-differential, these types of differentials were generally used by manufacturers in older, inexpensive applications. In addition to the (B13) Sentra SE-R, early Mazda Miatas used a viscous LSD before switching in 1994 to the superior Torsen design.

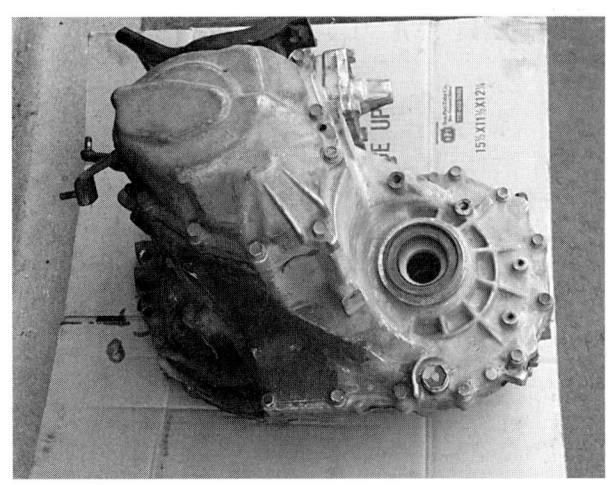

Torsen (Helical) LSD—The Torsen LSD gets its name from its function as a TORque SENsing piece of equipment. It works the same as an open-differential until one wheel begins to lose traction, at which point the gears bind, worm wheels interact, and torque, multiplied by the inherent torque bias ratio, or TBR, is channeled to the "least slipping" wheel. The TBR basically dictates the "locking effect" of the differential, or at what ratio the torque is transferred. A TBR of 4:1 would indicate that the differential is able to provide up to four times the amount of torque to one side and still stay locked, assuming that wheel can "use" that much torque.

Note: Both Torsen and Quaife LSDs are helical differentials, which simply means "shaped like a helix or spiral." The unit from Quaife is marketed as the Quaife ATB, which stands for Auto Torque Biasing, sounding, of course, very much like torque-sensing from Torsen.

Torsen LSDs react much more quickly and efficiently than viscous units, slipping less and transferring more—and more immediately. The Torsen LSD is extremely versatile, seeing duty in vehicles as diverse as the Hummer and the FD (1993–1995) Mazda RX-7. They are not rebuildable as clutch-type LSDs are, but they also do not suffer wear like a clutch LSD, either. A disadvantage to the Torsen LSD surfaces when wheel-lift occurs. Those cars that tend to lift a wheel in the air under hard cornering (Volkswagen GTI) effectively disable the ability of the Torsen LSD, which requires both wheels to have at least some grip to work properly (granted, it must be a lifted drive-wheel to have an effect, rather than the inside rear wheel on a front-drive Honda, for example). Torsen has also developed a technology called Torsen T-2R that benefits from preloading to help combat those circumstances where a wheel loses contact with the ground, though on most

Factory transmissions have one of two differential styles: an open differential, or some version of limited slip differential. This E153 (1993–1995 SW20) has a factory viscous LSD, which is verified with the axle removed by looking in that little hole: an LSD transmission will have a bar clearly visible. Photo courtesy Brian Hill.

LSDs, the higher the preload, the higher the drag. There's always a compromise.

Clutch-Type LSD—Very basically, clutch-type LSDs, such units as manufactured by Kaaz and Cusco, use an assortment of clutches and springs to resist being turned at different rates, without the action of viscous fluid. The stiffness of these springs dictates how much twisting force it requires overcome them. These LSDs are more suitable for drifting than Torsen units due to their durable and adjustable structure, though they also require rebuilding at intervals dictated by your use and the brand. They are also noisy and are not as smooth in their response as Torsen units, and must be broken in properly—Kaaz recommends a series of figure 8s for 30 minutes, for example.

The LSD Name Game—There are a variety of trade names involved in limited slip differential marketing, so much so that it can be confusing what does what, and what are the actual differences between units. Many American manufacturers use unique trading names as well, gimmicky stuff such as Safe-T-Track, Equa-Lock, and Traction Lock, as well as the old-school cool "posi-traction," which over the years simply became "posi," or "posi rear-end" during Wednesday night staging lane grunting at those south end drag strips. Honda went their own route with the Honda Prelude Type SH model to help alleviate understeer, developing a system called Active Torque Transfer System. This system is based, though, more upon computers than gears and clutches. There is even a "non-LSD" LSD, the unfortunately named Phantom Grip.

A less expensive alternative to the above LSDs, the Phantom Grip uses a handful of parts—basically a couple of plates providing force against stock spider gears—to provide an action similar to that of a clutch-type LSD, though I would not call it a limited slip differential any more than I'd call the Blitz defuser a true downpipe. The Phantom Grip is a somewhat newer product, and its simplicity is a wonderful thing. It has not been tested enough for a thorough evaluation under all circumstances, but for the price, up to a third of what a standard LSD costs, it may represent a good opportunity for you to get your hands dirty on your own, pulling your transmission and installing it yourself. Like the Megasquirt engine management system, the Phantom Grip is very much a hands-on, end-user defined experience, a learning tool being so relatively simple and affordable. Whether or not you want to risk differential issues when their install is such a hassle, though—and potentially expensive if you don't do your own wrenching—is up to you.

The Quaife ATB LSD, or Auto Torque Biasing limited slip differential, is a very nice unit, superior to viscous units in its ability to transfer torque, and much smoother in its activation than a clutch-type. Photo courtesy Quaife Engineering.

ADVANCED ENGINE PERFORMANCE
Lightweight Flywheels

The Idea: Reduced rotational mass for easier rev-matching and increased engine responsiveness

Any time you can reduce the mass the crankshaft is forced to rotate—and the flywheel is perched right on the end of the crankshaft—you will make it easier for the engine to do work. We (should) all seek out ways to make our MR2s lighter—see the ZZW30 for a demonstration of the benefits of "adding lightness."

There have been formulas floated about how much losing X amount of weight will affect performance—generally the theory is for every 100 lbs. you lose, your quarter-mile E.T. should drop 0.1 second. There are similar ratios attempting to calibrate the difference in losing "rotational mass," and weight from the trunk for example. *Sport Compact Car* magazine has suggested that every 1 lb. of rotational mass is like 15 lbs. of static mass—that is, the mass in your trunk or engine bay mass that doesn't move. No matter the exact formula, it is clear that reducing flywheel weight is a good thing. Among the enhancements to the all-conquering Acura Integra Type-R was a lightweight flywheel, which is testament to its success as a modification—that car did very little wrong.

The most significant impact a lightweight flywheel will make on your MR2 comes courtesy of its newfound decreased inertia: engine responsiveness is greatly enhanced, offering up improved feel, and improved gear change performance as well. Blipping the throttle is made easier and more immediate. This allows you to

Advanced MR2 Performance Modifications & Theory

The flywheel is bolted directly to the end of the crankshaft, as shown on this 3S-GTE. Make it lighter, and you're putting less load on the crankshaft. This is a stock 3S-GTE flywheel that has been machined and lightened—not a process I recommend. Buy a billet steel or aluminum lightweight flywheel.

Aftermarket head gaskets like this example from HKS come in a variety of thicknesses to provide an array of compression ratios, as well as unique sizes to accommodate different bore sizes as well.

more quickly access different points in the rev-range to more easily enable rev-matching, especially when downshifting. Coming into a corner in third gear and knowing that you'll need second, a lightweight flywheel will allow you to more precisely pick your rpm for rev matching.

So if a lightweight flywheel is so great, why didn't your MR2 come with one from the factory? As with most cases, it is a matter of compromise. OEMs often make flywheels heavier for increased driveability. Heavier flywheels are like energy sinks, making launching your MR2—whether aggressively or leisurely from a stop sign—smoother, and more effortless. And while, for a data point, the stock AW16 flywheel is already fairly light at around 13 lbs., the Chrysler Conquest/Mitsubishi Starion of the 1980s had a 32-lb. flywheel!

Do Not Lighten A Stock Flywheel—Whatever you do, *do not* simply lighten your stock flywheel by machining off some material. The lightening process can compromise the structural integrity of the unit, leading to catastrophic failure under severe load. Considering the flywheel's proximity to your backside in an MR2, it is not worth the gamble. Both steel and aluminum are used by manufacturers to form new billet units safely. This is not an area to save money.

Daily Driving and Lightweight Flywheels—Ultimately a lightweight flywheel is more of a seat-of-the-pants modification, enhancing what the car feels like when you drive it, versus a modification offering stunning performance.

It isn't so much a hardcore performance upgrade providing immediate, unmistakable gains as it is a complement to an overall more comprehensive package and buildup plan, i.e., a clutch-lightweight flywheel package combination. In other words, don't ever pull a healthy clutch just to install a lightweight flywheel.

A common criticism of lightweight flywheels is that they cause a car to be difficult to launch—even at a poke's pace while commuting. While this could be true for an extremely light, race-only flywheel, overall the type of clutch itself has far more potential to injure the driveability than flywheel weight. Once installed, your average 9 lb. Fidanza flywheel should be easily adjusted to in daily commuting in your MR2.

Performance Head Gaskets

The Idea: Proper sealing between the head and the block

The head gasket is an infamous character in tuning, kind of like a referee—only mentioned when they fail. Generally, if your head gasket fails an entire rebuild is called for, and when this happens, people are generally resolved to not have it fail again.

The smart modifications are adjustments—so what are we adjusting here? What is the stock piece not doing that we need it to?

Right away, most aftermarket head gaskets are made of various types of metal, usually steel, and so are stronger than factory head gaskets. Secondly, aftermarket head gaskets are available in a variety of thicknesses. Since the head gasket sits between the head and the block, the thickness of the gasket directly affects the compression ratio of the engine: a gasket thicker than stock will reduce the

Toyota MR2 Performance

Absolutely critical to the success of any head gasket, stock or aftermarket, is proper sealing between the head and the block. Precisely flat surfaces on both sides are requires for correct mating. 3S-GTE head shown.

compression ratio, while a head gasket thinner than stock will raise the compression ratio. They are also available to accommodate a range of unique bore sizes, and this kind of adjustability makes the aftermarket head gasket the domain of the MR2 owner rebuilding an engine with performance in mind. Stock head gaskets, while Toyota-strong, do not offer this adjustability.

Stock Head Gaskets vs. Aftermarket—What about the average stock 4A-GE, 4A-GZE, or 3S-GTE? Even one pushing 100 hp or more over stock?

The proper sealing of the gasket between the block and head is far more critical than the material the gasket is made of. Even the proper torqueing of the head bolts, and the condition of the bolts themselves contribute significantly to the proper sealing of the head to the block. Gasket material matters, but is easily 3rd or 4th in priority.

OEM Toyota head gaskets are appropriate on all but the most ambitious buildups, at which point any number of options become available. Many MR2 owners have had their stock Toyota gaskets survive 25+ psi of boost for years before failing.

Honestly, this has more to do with the glue-like substance old gaskets that have been heated and cooled thousands of times turn to than the inherent strength of the OEM gasket, but it is strong enough that it should be clearly understood that the stock gasket, at the very least, is not a weak point, 4A-GE, or 3S-GTE.

Note: The 4A-GZE head gasket is less durable than that of the 3S-GTE, with head gasket failure-rate on the 4A-GZE higher than the 3S-GTE, but, again, how durable your head gasket proves to be depends on a variety of factors, including whether or not its ever been overheated, and relative mileage.

TTE and HKS Head Gaskets—For those 3S-GTE owners that want OEM quality with the added strength of a metal composition, a stock gasket that represents a wonderful choice is the TTE gasket, or the "Team Toyota Europe" gasket. This gasket, much like the gasket installed on the Gen III 3S-GTE stock, is a factory metal head gasket, and being OEM, can be trusted to possess sufficient engineering.

HKS manufactures another popular 3S-GTE metal head gasket, and for the 4A-GE as well. Being that they have input into how close the valves get to the pistons, head gaskets of varying thickness can also be used to roughly achieve a desired compression ratio, though ultimately that is best left to custom pistons and head-porting specifications.

Copper Head Gaskets—Copper head gaskets are another option. Copper head gaskets are often used on highly boosted engines because, when properly installed, they tend to be able to withstand the pressure inherent in these setups better than multi-layered steel gaskets or composite gaskets.

Some tuners have gone the extra step to O-ring the head with copper O-rings. This is a process where copper rings are glued or otherwise fixed around each bore to provide a seal. The entire deck of the block can be O-ringed as well. As with other head gaskets, they can also be had with varying thicknesses to render desired compression ratios. The problem with copper gaskets and O-rings it that they tend not to seal as well as other gasket styles because, unlike composite gaskets, they do not compress when the head is installed. There are companies that now manufacture rubber-coated head gaskets to help solve this problem, though considering the financial consequences of a head gasket failure, unless you're building a monster, there is little reason to experiment with a copper head gasket.

Intake Manifolds

The Idea: To distribute air evenly to the four combustion chambers

Unless you're looking to double factory horsepower ratings, the AW11 (4A-GE and 4AGZE), and the ZZW30 (1ZZ-FE) are not restricted by their intake manifold performance, and if there is no restriction, look elsewhere to produce horsepower.

The tremendous horsepower potential of the SW20 (3S-GTE), though, make its manifold design actually an issue for many tuners. In fact, many owners have been known to treat their intake

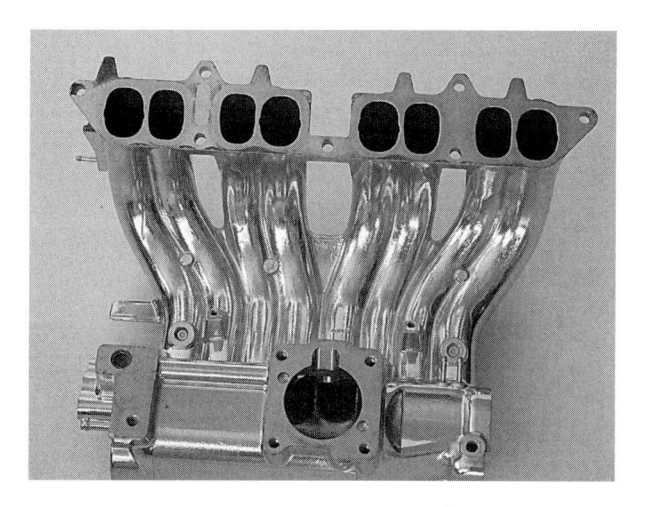

ADVANCED MR2 PERFORMANCE MODIFICATIONS & THEORY

The job of the intake manifold is simple: to evenly distribute air into the cylinder head ports. This Gen II 3S-GTE intake manifold can be identified by its (TVIS) dual-runner design.

For the third-generation 3S-GTE, Toyota reduced the number of runners on the intake manifold in half, and reduced the size of the intake ports to increase intake velocity. Photo courtesy Jensen Lum.

manifolds to the Extrude honing process even at stock power levels, simply to ensure each runner is feeding the combustion chamber downstream the same amount of air as the one beside it.

For years the power-hungry MR2 owner has had to live with the stock intake manifold— and for the vast majority of MR2 owners, this has not been a problem. Replacing an intake manifold is not fail-safe like slapping a larger exhaust on a turbocharged car; there is a lot to consider with an intake manifold, not to mention very few alternatives on the market or data to research what is out there. Bigger is not always better.

Performance automotive modification should always be thought of in terms of well-considered adjustments to address some known deficiency—the car understeering, the brakes fading with aggressive use, or a "bottleneck" in the air intake system. When you start doubling airflow through the stock intake manifold through radical boost increases, it is logical to assume that eventually that manifold will become inadequate. At that point there is cause for empirical data to help identify what exactly about the stock manifold requires "adjustment," and what those adjustments should be.

Toyota Variable Intake System (TVIS)

Like the large-port 4A-GE, the stock Gen II 3S-GTE intake manifold features eight individual runners, four of which are blocked by a vacuum-actuated butterfly plate off at low rpm to increase low-end performance. While worth keeping on a stock 4A-GE, how useful TVIS is on the 200 hp 3S-GTE that is being force-fed air to begin with is of some question to many MR2 tuners. It does seem to increase low rpm air ingestion, and there is at least some evidence that Toyota may have designed TVIS-actuation to help suppress detonation.

The stock 3S-GTE also features a very small plenum—the "collecting chamber" immediately after the throttle body. When being forced air from a turbocharger capable of flowing significantly increased cfms over the stock turbocharger at stock boost, a larger plenum is ideal. It is interesting to note that in its own evolution of the 3S-GTE, the Gen III 3S-GTE, Toyota abandoned both the TVIS, 8-runner design, and enlarged the plenum as well.

Aftermarket Intake Manifold Design—In the design or modification of any intake manifold, two of the main factors which must be considered are runner length, and plenum sizing and design. The stock 3S-GTE manifold has repeatedly been measured to flow unequally through these runners, feeding the middle cylinder head ports, especially the #3 cylinder, more air than the others, creating a lean condition–meaning there was less fuel than air—which can lead to decreased performance and at worst, major engine damage. For years, this was thought to be the reason why many 3S-GTEs failed, but when 3S-GTEs with new or modified stock intake manifolds also failed, the real problem—the less interesting concept of proper tuning of fuel and timing—was identified as the culprit.

Though JUN manufactured an aftermarket plenum used with the stock runners, underscoring the critical nature of the plenum's design, Ross Machine Racing was one of the first and few companies to step up and design an entire intake manifold for the 3SGTE. But that runner length issue keeps making things complicated.

67

Toyota MR2 Performance

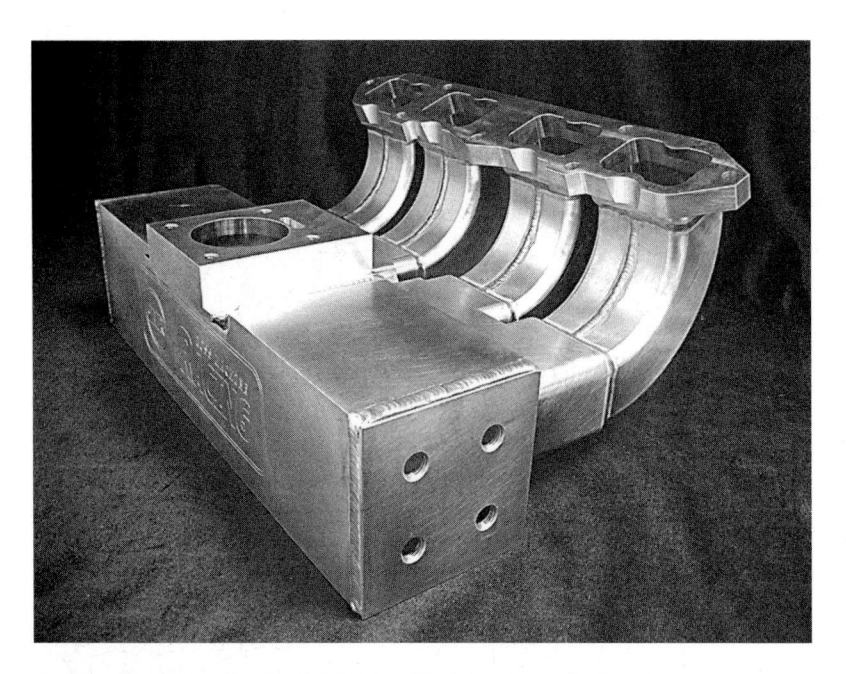

The Ross Machine Racing (RMR) 3S-GTE intake manifold reduces runner length, which has the potential to increase upper rpm performance, at the cost of low rpm grunt. It also increases the plenum size and alters its shape. Photo courtesy Ross Machine Racing.

To put it very simply, shortening the runner length will generally shift the powerband higher (to the "right" on a dyno-plot). The Acura Integra Type R featured shorter intake manifold runners than the Integra GS-R, and in combination with cylinder head porting and aggressive camshaft design resulted in a higher redline, and more horsepower near that redline than in the GS-R.

As mentioned, the intake manifold from Ross Machine Racing features runners that are shorter than factory, which by itself should increase high-rpm performance, theoretically at the expense of low-rpm torque. After production of this manifold was finished and grassroots-level testing ensued, two things became clear: 1) It did indeed underperform the stock manifold at everyday driving rpms, under 5000 rpm, and 2) At elevated rpms, it made considerably more power. So the question comes: are these gains in performance worth the considerable expense and labor? And if so, which manifold is "best"? Unfortunately, it depends.

On stock engines with stock turbochargers, an aftermarket intake manifold makes little sense. A schedule for replacing the 3S-GTE intake manifold is available in Chapter 6, but overall choosing exactly when it is ideal to replace the factory intake manifold, and with what, depends on everything from your intercooler setup—trunk or side, air or water—to the high-performance application you most often ask your MR2 to perform in—road course, drag, street, etc. It also depends on the torque-curve you like, and the turbocharger you have in mind to pair it with. Remember, an engine is a system, not a single thing, and all components must be thought of in relation to one another more so than simply choosing individual "best" parts in a piece-meal fashion.

Extrude Honing

Extrude honing is a process by which very coarse, abrasive putty is forced through cavities like intake manifold runners, mildly porting them for increased airflow and, in the case of the 3S-GTE stock intake manifold, helping equalize the airflow across the cylinders. It can be considered a compromise between keeping your stock manifold, and getting a big aftermarket unit like the Ross Machine Racing manifold. The extrude honing process can also be applied to exhaust manifolds, and literally anything else that needs honing from the inside.

In spite of what would seem common sense, some tuners can make obscene amounts of power with the stock intake manifold, while others set about extrude honing or replacing their stock manifold the minute they replace the turbocharger. Certainly there is power to be made enhancing how we get the air from the turbocharger to the intake valve in the cylinder head, and the intake manifold is a big part of that air distribution process.

We make specific recommendations in our staged buildup chapter, but bear in mind that keeping things simple is almost never a bad idea, and that there are easier ways to make horsepower, especially on a turbocharged car, than fudging with your intake manifold. This reality makes the extrude-honed stock intake manifold an option worthy of your close consideration, and is personally what I used on a 300 whp MR2 Turbo of mine.

Throttle Bodies

The factory Gen II 3S-GTE throttle body is 55mm, the Gen III 60mm. In big horsepower setups, throttle-bodies as big as 75mm are used, and sourcing throttle-bodies this large, with appropriate throttle-position sensor, vacuum lines and other considerations can be tricky. In the SW2x, we are talking about a relatively lightly modified car that has been out of production for nearly fifteen years!

Ideally, you'll have a throttle body that simply does not bottleneck the flow of air into the intake manifold. You will also need to port match your intake manifold after having installed an aftermarket throttle body. Port matching is a

Advanced MR2 Performance Modifications & Theory

The throttle body is basically a gateway, controlling how much air is ingested by the engine through a butterfly plate actuated via the throttle. JDM Gen II 3S-GTE shown. Photo courtesy Mike Choi.

This AW11 has an 3S-GTE stuffed in it, and it uses an air intake and intercooler configuration similar to the SW20, so factory throttle body placement worked well. The simpler, the better.

The 4A-GZE has a unique air-intake system that features very direct plumbing due to the intercoolers proximity to the air intake and the intake manifold. It's all right there. Photo courtesy David Jones.

This highly tuned 3S-GTE uses an air-to-water intercooler, which also necessitated the owner to move the throttle body to the driver's side of the engine bay—and some creative methods of manipulating the throttle-plate. Photo courtesy Brian Hall.

process by which matching and sealing ports are matched via a metal grinder of sorts to ensure a 60mm throttle body isn't bolted to a 57mm "hole" in the intake manifold (which would completely defeat the purpose of the throttle body; this concept also applies to exhaust system components—3" downpipes into 2.5" b-pipes, exiting into 2.75" muffler canisters don't make a lot of sense).

As horsepower increases 25%–35% over factory ratings, details like throttle body sizing (and corresponding port-matching) become more critical. At factory power levels, modifications like the throttle body can be more trouble than they're worth.

Throttle Body Placement—Beyond runner length and volume, and plenum sizing, another consideration of any aftermarket intake manifold is throttle body placement. Most manifolds, like the Ross Machine Racing manifold for example, allow the buyer the option of where they want the throttle body placed—on either end, or in the middle as it is on the factory setup.

Why move it? Sometimes certain turbocharger and intercooler arrangements require innovative plumbing methods, and moving where the throttle body is can help.

A trunk-mounted intercooler, for example, might benefit more from the throttle body being mounted to one side to eliminate wasteful bends in the intercooler piping. Intercooling a highly tuned, mid-engine, forced-induction car can create

Toyota MR2 Performance

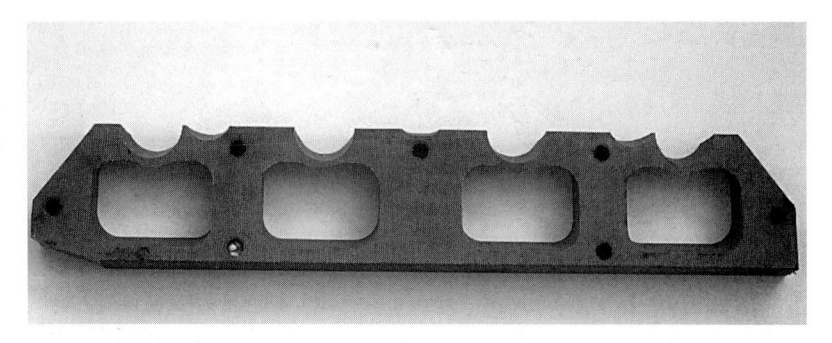

A hot intake manifold will obviously heat the air it encounters. While engine bay temperatures can heat the manifold, a significant source of heat is the engine block itself. A phenolic spacer provides a physical thermal barrier between the engine block and the intake manifold.

There are two ways to increase the displacement of the engine: increase the bore size or lengthen the stroke.

unforeseen complexities, and throttle body placement is one of them.

Phenolic Spacers

A phenolic spacer is a simple and effective performance bolt-on. In the case of the 3S-GTE, a phenolic spacer is made to exactly duplicate the shape of the TVIS butterfly plate, but is made of resin. This resin plate basically creates a heat barrier between the cylinder head and the intake manifold, preventing the engine block/exhaust manifold from continuously heating the manifold, and thus the air inside.

We spend so much effort cooling the intake charge with intercoolers, water injection and so on, it only makes sense to not turn around and reheat the air we just bent over backwards to cool. Improved detonation resistance and denser intake charges both result from the reduction in intake temperatures. Notice the performance difference in your car on an 80°F day versus 60°F? That is well within the expected performance range of a phenolic manifold spacer. Recent testing on these spacers by *Autospeed* revealed intake temperature reductions no fewer than 20°F, and reductions by as many as 60°F, with the spacers also allowing the manifold to initially stay cool much longer than the factory manifold as well.

Installation is moderately difficult which is why it is in the advanced chapter, mostly due to the location of the plate, one of the few strikes against it as a modification. But with proper tools, a person of average mechanical skills should be able to complete the install in 2 to 4 hours.

TVIS and Phenolic Spacers—Due to Toyota's unique TVIS system on the large-port 4A-GE and the Gen II 3S-GTE, which is mechanically part of the intake manifold, installing a phenolic spacer is a simpler alternative to gutting the factory TVIS plate (which amounts to removing the butterfly plates and sealing the holes). If you've planned on removing your TVIS, this is a point to consider. If you don't have—or want—functioning TVIS, and the intake manifold is going to be off anyway, there is no reason to skip this modification.

Stroker Kits

The Idea: Increased displacement and power production across the rev range

We could spend several pages discussing whether or not that is really true—the whole domestic vs. import "is there a replacement for displacement thing," but for now we'll simply say that appropriately increasing the displacement of an engine can yield increased torque and overall engine output, so if that boosted 2.0 liter 3S-GTE is producing 380 hp, all else being equal, increasing the displacement will generally have a positive effect on the torque output across the rev range.

By increasing the displacement, you are also making it more potent and receptive to other power-increasing changes. Every stock MR2 was 2.2 liters or less, and every forced-induction, high-output MR2 2.0 liters or less. Aside from bolting on a fast-spooling moderately sized turbo, increasing the displacement can significantly liven up what happens on the lower end of the rev-range, and ultimately offers tremendous returns on everything your engine is asked to do.

Stroker kits are basically resized internal engine components that effectively increase the displacement of your engine. These kits have become increasingly popular over the last several years as the MR2 tuning community has sought out ways of spooling larger turbochargers, and improving powerbands that have no place in a mid-engine sports car.

The mass-market stroker kits for the 3S-GTE,

ADVANCED MR2 PERFORMANCE MODIFICATIONS & THEORY

The main components necessary to complete the stroking of your MR2's engine are the crankshaft, pistons rods, and pistons. Aftermarket forged pistons like this piston from JE can be ordered with a custom compression-ratio as well.

such as those available from JUN and ATS Racing, increase the displacement 10%, in this case 0.2 liters, rendering the same displacement of the normally aspirated 5S-FE, but with a turbocharger to boot.

The Components of a Stroker Kit—There are effectively two ways to increase the displacement of an engine: increase the bore of the cylinder, or increase the stroke of the rod and piston.

The stroke is the physical distance the piston rod moves from bottom-dead center (BDC)—its lowest point—to top-dead center (TDC), or its highest point. The physics behind this process are fairly simple: a "stroked" crank and connecting rod will generate more force, or leverage, than a shorter rod.

Stroking is less labor-intensive than boring and honing the cylinder walls—another method of increasing displacement—and is by far the more common stroking method in the small-displacement aftermarket. Many smaller-displacement engines do not have enough material between the cylinders for large machining, and so an increased stroke is the most common method of displacement increase for MR2 owners.

Rod Ratio—Rod ratio plays a significant role in the success of a stroker engine, and in the character of the engine as well. Normally MR2 owners needn't concern themselves with this ratio; Toyota did all the work for you. However, if you are seeking to increase the displacement of your engine, suddenly you're in the market for a little rod ratio information.

rod ratio = rod length ÷ stroke

The engine's rpm tells you how many times the crankshaft is turning per minute, but actual piston speed is determined by the rod length and stroke.

3S-GTE STROKER KITS AND TOYOTA CAMRYS

Luckily enough for MR2 owners, many Toyota parts interchange, and work cooperatively. The crank from the 2.2 liter normally aspirated MkII, and older Toyota Camrys, can be used in the 3S-GTE block to render a longer 91mm stroke (from 86mm stock), which is exactly what ATS Racing does in their kit.

The 3S-GTE cranks has smaller journals than the crank from the 5S-FE, so the rod journals from the 5S-FE crank are machined to the diameter of the 3S-GTE journals. This ensures a wider selection of rods available since 3S-GTE rods, stock or aftermarket, will fit. One might think that the crank from a 135 hp engine would not be ideal in engines producing 3-4 times that, but Toyota typically overbuilds everything, and the 5S-FE crank is an excellent example of that. No wonder Camrys run forever.

Note: Since we're using a Camry crank, to keep from having to use a Camry clutch, something other than a Camry flywheel must be sourced. Interestingly enough, the flywheel from the ST165 Toyota Celica All-Trac does the trick. This flywheel has the same bolt pattern as the 5S-FE crank, but is as large as the 3S-GTE flywheel (the ST165 received the Gen I version of the 3S-GTE, as a matter of fact).

The rod is obviously physically connected to the crank, and as the crank twists, stroke already having been established, the speed and accelerative forces imposed on the piston are set by math.

Additionally, low rod ratios mean high rod angles. Since the piston rod effectively is "moved" through a circular path based on the movement of the crankshaft, a low rod ratio will place a short piston rod at odd angles, namely more into the side of the combustion chamber walls than straight up. Lengthening the rod length will correct this problem, reduce wear in the combustion chamber and on piston rings, while also decreasing friction from the movement of the pistons—always a good thing. This requires changes elsewhere though, such as the height of the piston and the crankshaft itself.

The acceleration of the piston leads to significant stress on the bottom end—crankshaft, piston rod, pistons, etc. As these forces increase, the physical ability of the engine to literally hold together can be exceeded. The relatively long stroke of the B18C5 in the Integra Type R rendered piston speeds at its 8400 rpm redline approaching that of Formula 1 engines—engines that can see 20,000 or more rpm. As a point of data from another manufacturer, Honda utilized the same basic engine in the 1.6L B16 as they did in the 1.8L B18, each engine basically being a stroked (or de-stroked) version of one another. Though the 1.8L has a 0.2L displacement advantage over the B16, most enthusiasts agree the B16 feels livelier, and more eager to rev, this subjective feel mainly courtesy of its superior rod ratio.

In addition, rod ratio also renders dwell-time of the piston at top-dead-center (TDC), which can improve combustion performance, and a variety of other factors that all contribute the volumetric efficiency—how efficiently your MR2's engine turns air and fuel into horsepower—of your engine. Not easy stuff to change without compromising. The point here is that it's not all about displacement. Rod ratio matters.

Interestingly enough, a common practice, especially for high-speed runs where an engine operates in a very narrow rpm window, is to actually de-stroke an engine—that is, reducing the displacement by reducing the stroke—in the name of increasing high-rpm performance.

Oversquare vs Undersquare—An engine featuring a bore that is larger than the stroke is said to be oversquare, while an engine with a stroker longer than the bore is said to be undersquare. Generally speaking, oversquare engines are more eager to rev to higher rpms. The 3S-GTE is actually perfectly square, its bore and stroke both an even 86.0mm (the main factor to the 3S-GTE's tepid 6000+ performance lies in the tiny exhaust wheel/housing on the stock CT26 turbocharger—a proper turbocharger will send the 3S-GTE screaming).

3S-GTE Retail Stroker Kits—Through the wonder of physics, the added length to the stroke adds displacement—displacement that can help spool larger turbochargers, and increase the torque band everywhere. Most 3S-GTE stroker kits increase the displacement from 2.0L to 2.2L, an increase of 10%.

Two of the most noteworthy 3S-GTE stroker kits are the kits made by JUN and ATS Racing. Of course everything JUN touches oozes quality, but that quality costs, in this case rendering a prohibitively expensive stroker kit that, ultimately, gives you the same extra 0.2 of displacement that the ATS kit gives you, minus the customer service and technical support the Texas-based ATS Racing can offer.

JUN worked with Cosworth, a legendary engine builder based in Northampton, England, to engineer a stroker kit for the 3S-GTE. Cosworth has decades of experience developing engines for Formula 1, CART, Indycars, and the World Rally Championship as well. Most of JUN's products, while often prohibitively expensive, are well-researched and thoroughly engineered, while the ATS Racing stroker kit has the MR2-community-proven performance behind it.

ADVANCED INTERCOOLING PERFORMANCE

The Idea: Letting the heat from compressed air escape as efficiently as possible

The the most effective forced-induction setups will generally be intercooled. Toyota must have agreed, because they equipped all forced-induction MR2s with air-to-air intercoolers (though they did use an air-to-water intercooler in the front-engine Celica All-Trac).

Air temperature and pressure want to equalize, so heat will escape "into" the cooler air around it if you let it. The supercharger available on the AW16 and the turbocharger available on the SW20 compress the intake charge with great force, creating boost. Each of these processes generates tremendous amounts of thermal energy—energy that needs to be dissipated somehow.

Much like an exhaust, sway bar, or aftermarket turbocharger, we're basically looking to increase the size here, specifically an aftermarket intercooler with a core thicker than stock, while minimizing pressure drop.

Pressure drop is basically the difference in boost pressure as it exits the turbocharger, and that which actually makes it to the intake valves. The most effective intercoolers will provide the coolest intake charge with the least amount of pressure drop, all other things being equal.

Spearco and GReddy manufacture aftermarket sidemount intercooler kits for the SW20, and many speed shops, like KO Racing and AutoLab, build

Compressing air gets it very hot in a hurry. The intercooler's job is basically to let that heat escape, while minimizing the pressure drop. A stock 4A-GZE intercooler is shown.

Because of their considerably higher output, and generally more aggressive-tuning owners, SW20s see intercooler upgrades far more often than AW16s, but the physics remain: to consistently reduce the intake charge and resist heat soak, a larger, more efficient intercooler is necessary. Photo courtesy David Jones.

their own kits using cores from intercooler manufacturers, basically adding end tanks and intercooling piping to make it a bolt-on affair.

The AW16 has fewer—if any—bolt-on aftermarket intercooler kits, but building your own isn't rocket science. It basically boils down to ensuring an intercooler will physically fit and can effectively be secured, and then figuring out how to plumb it, something your average speed shop can do for reasonable cost.

Intercooler Performance Factors—The biggest priority from an intercooler should be consistency; the more consistent the intake charge, the more precise tuning can be. Tuning for an intercooler that isn't able to consistently reduce the intake charge is an uncertain—and dangerous—practice. Along with cooling potential, the ability to flow high cfms, and provide the least possible pressure drop while cooling the charge air, the intercooler's ability to resist heat soak should be your highest priority on a road-going, performance-tuned forced-induction MR2.

The MR2, by its mid-engine design, suffers from relatively poor airflow throughout the engine compartment compared to other engine and chassis configurations, which isn't much of an issue stock, but can become one when factory boost levels are exceeded. Toyota engineers actually did a very commendable job getting air over, through, and around the engine bay, and many of the design features that you see—or don't see—when you pop your engine lid were created to do exactly that.

Engine bay temperatures rarely climb above 25–35°F above ambient while the car is moving in normal driving. Still, when so many of our forced-induction counterparts have access to the front bumper or hood for air (Mitsubishi Evo) or a giant hood scoop (Subaru WRX), MR2 owners have to often get creative in their method of intercooling. Custom air-to-water setups, "trunk-mounted" intercoolers, and even top-mounted setups similar to the one utilized by Subaru in the WRX, have all been put into duty by MR2 owners seeking a cooler, quicker-recovering intake charge.

Note: A good rule of thumb is to somehow supplement your factory intercooling by the time you've increased your power output more than 10%. In fact, even for bone-stock engines, though the overall horsepower gain is minimal, the performance consistency increase due to heat-soak resistance make the investment worth it.

Supplementing the Sidemount Intercooler— Intercoolers are made of metal, which quickly absorbs and retains heat. Heat soak occurs when an intercooler is overcome with heat, and is unable to dissipate any further heat energy. If the intercooler cannot dissipate heat, intake temperatures will rise dramatically, and the engine management system will be forced to take steps to prevent engine damage, namely reducing timing. In such a state, your MR2 will feel sluggish, and should not be

The Spearco sidemount intercooler for the SW20 is a bar and plate type intercooler, featuring dimensions 6" x 5.9" x 11.75", bigger than the 4" x 7.5" x 9.5" sidemount kit from GReddy, and superior in cfm flow-test numbers to the GReddy unit as well. Still, either intercooler is fairly comparable. Ultimately they're both limited due to their sidemount design.

Toyota MR2 Performance

Aftermarket fans like those from Spal are rated by cfm—their ability to move air. Upgrading the stock intercooler fan is an inexpensive way to modestly increase the performance of a sidemount intercooler, stock or aftermarket. When used in conjunction with a shroud, its performance can be increased even further.

driven aggressively, to prevent detonation.

So what do you do? If your intercooler, stock or aftermarket, heat soaks—and it will if you're doing back-to-back wide-open throttle runs—one way to address this issue is just to drive without allowing the engine to generate boost. The airflow through the intercooler will naturally cool it eventually. But are there more proactive solutions? Aftermarket sidemount intercoolers are usually larger (more surface area to release more heat) and more efficient than the stock unit, allowing for better cooling of intake charges, especially those generated by higher boost levels. But even these units simply delay the onset of heat soak longer. There are methods of supplementing the heat-shedding potential of this improved unit.

The easiest way to improve the efficiency of a sidemount air-to-air intercooler is by forcing air through the intercooler with a fan. The stock intercooler uses a fan as well, but is not turned on until engine bay temperatures reach 140°F. Being fed by the passenger-side vent on the MR2 Turbo, any intercooler stuck there will undoubtedly be fed fresh air, but a 7.5" fan by Spal can move up to 440cfm (cubic feet per minute), and there are options out there to house 9" or bigger fans as well. Typically housing this fan is a shroud. Intercooler shrouds are bolted to one side of the intercooler, anywhere from 1" to 2" away from the intercooler surface, increasing the efficiency of the fan, and the intercooler itself.

Beyond a fan and shroud, we have a third option that can be used in addition: an IC sprayer of some sort—whether with water, as the Mitsubishi Evo and Subaru WRX STI employ, or one slightly more expensive that soaks the intercooler with extremely cold carbon dioxide or nitrous oxide.

Limits of the Sidemount Intercooler—Beyond airflow through the engine bay, the main restriction of a sidemount intercooler is space. There's simply not much room for a sufficiently large intercooler. Even at stock horsepower on a 3S-GTE, consistent cooling of the intake charge is questionable at best. Back-to-back runs at stock boost with a stock turbo will begin to overwhelm the stock intercooler.

Increasing the size—mainly the thickness of the intercooler's core—will certainly help, but bear in mind the limitations of even this modified setup. A sidemount intercooler on the SW20, no matter how thick, has its restrictions. At levels over 300 whp, even a thick aftermarket core from Spearco, for example, is no longer sufficient, and will heat soak fairly quickly.

If you're planning on staying below 300 whp, the sidemount intercooler can be made to work with a few tweaks. If you plan on consistently making more horsepower than that, you need to consider other designs.

Water Injection

The Idea: Supplemental intercooling offering combustion chamber steam-cleaning and increased detonation control

Water injection has been used for decades, mainly to suppress detonation. Detonation is basically the uncontrolled combustion of the air/fuel mixtures, usually caused by excessive compression and heat. Roughly speaking, water injection impacts detonation most greatly through additional cooling. The short of it has to do with water's natural latent heat of vaporization.

What is Water Doing in the Combustion Chamber?—For one gram of water, the amount of heat energy required for vaporization is 540 calories at a temperature of 100°C. In layman's terms, it basically means that when we introduce water as a mist into the hot intake charge barreling straight for the intake manifold, that water will turn to steam.

An intercooler shroud is a lightweight, aluminum piece that shields the "hot side" of the intercooler from the engine's heat, and increases the efficiency of the fan as well.

ADVANCED MR2 PERFORMANCE MODIFICATIONS & THEORY

ENGINE LID FANS

Many owners use fans on the engine lid to help dissipate heat from the engine. The industrious MR2 aftermarket has developed several engine lid shrouds to house such fans, oftentimes in stunning, polished metal that looks fantastic. These can be tempting.

In reality, these are only helpful during extended idling periods. While moving, the design of the engine bay and lid are adequate to evacuate heat out of the engine bay, and unless you are consistently monitoring engine bay temperatures in the name of the scientific method, don't assume your engine bay is indeed cooler—and thus better off—with the fans. (This is not even to mention that unless they're overheating, engines run more efficiently the hotter they run, not cooler.)

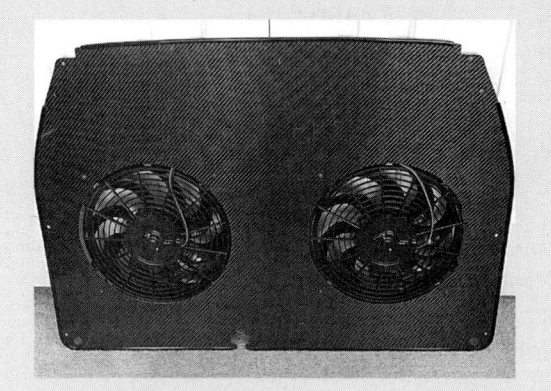

This engine lid shroud and fan setup is carbon fiber, and wonderfully engineered. It is also is very expensive, and more of a luxury than a real necessity, mainly providing lower engine bay temperatures during extended idling periods than offering a crucial overall cooling of the engine bay.

Both Jeff Fazio and David Vespremi's highly tuned SW20s use water injection for its many benefits. Whether you tune with it for horsepower or simply use it for durability's sake, it's undoubtedly effective. Photo courtesy David Vespremi.

That state change, from a liquid into a gas, requires a tremendous amount of energy, and energy is heat.

The amount of heat energy drawn from the air is significant enough to cool down the intake charge considerably, generally 50–75°F, though the actual amount in your application depends on other factors. What this means for your forced-induction MR2, beyond an obviously cooler intake charge, is detonation control.

Detonation control is what it sounds like—an enhanced ability of the combustion process to happen at the right time, when the piston and valves are expecting it. This added resistance to detonation can allow timing to be advanced beyond levels that would've been unsafe without it, gaining horsepower, and increased turbo spool. This benefit is very similar to that of high-octane fuel.

Detonation not only affects pistons, but can contribute to bearing failure as well. The physical forces of untimed combustion events are tremendous, much like the effects of a staggering increase in torque.

As we know, when torque increases, loads on rods and rod bearings increase, and any oil starvation at all—whether through incorrect viscosity, oil suffering through too many miles and associated thermal breakdown or simply via low quality—put a huge strain on the bearings, and over the long haul, can lead to premature rod bearing failure even if your engine never fails.

Steam Cleaning—A secondary, but possibly more critical, benefit comes with the reduction of hot spots in the combustion chamber by the gradual washing away of carbon buildup. These carbon deposits on the piston tops heat and cool at different rates than the surrounding metal. If they get hot enough, they act like little spark plugs, and basically ignite the air/fuel mixture prematurely, otherwise known as detonation, which is not a good thing.

These carbon deposits can also be removed through extremely careful, metered addition of water, siphoning up a bowl of water through a vacuum hose near the throttle body for example, but this is obviously a very dangerous process, so I'm not even going to outline it for the casual tuner. Too much water and your rods will bend like a banana—not a pretty sight. Just buy a water injection system, and enjoy all of its varied benefits.

Like an aftermarket intercooler, by itself water

Race fuel is wonderful for its high-octane detonation suppression, but it is not ideal to run consistently in a street-driven MR2. Water injection has a similar effect on detonation as race fuel, and is considerably less expensive in the long run. Photo courtesy Jeff Fazio.

injection does not add the huge horsepower numbers. It basically is a second intercooler. The best way, though, to think of it, aside from the consistent steam cleaning on your engine, is like relatively cheap race fuel. Those with standalone EMSs and other similar devices can "tune" with water injection, that is advance timing and boost to levels that would have been unsafe without the detonation suppression water injection offers. Certainly this is an option, and power increases of 10% or more are not out of the question, so mild 4A-GZE builds might expect 12-16 hp, and mild 3S-GTE builds 20-25 hp. As with everything else, as power levels increase, so do potential gains.

Water Injection Fail-Safes—Critics of water injection are not basically most concerned with the "independence" of the water injection system, components, water reservoir, etc. Once the factory ECU loses control over aspects of the combustion process, the slack in the chain is gone, and failings across systems can be catastrophic. There is simply less room for compromise, and the whole idea behind "for performance" water injection is to increase boost and/or ignition timing to levels unsafe without it; what happens if the water runs out? In these cases, without a fail-safe, if tuned to the ragged edge, instead of power simply falling, or a check-engine light clicking on as it might stock, the engine can self-destruct. Tuning for something that can run out—water in reservoir stuck out-of-mind in the trunk for example—is certainly something to be considered. This is where fail-safes come in.

Fail-safes are systems to help protect for less-than-ideal and unexpected circumstances, like running out of water at wide-open throttle and 18 psi of boost.

Many systems have such fail-safes built in.

Aquamist's 2D, for example, has a water fault output that detects that water is present and contributing to the in-cylinder cooling. If it notices water "missing," it can react in user-input ways, from cutting boost levels to alerting the fuel system to inject additional fuel to compensate for the missing water until the water can be replaced.

On simpler, more common water injection system like the Aquamist 1S, flow-monitoring devices can be added to warn the user/system when water is not available, whether because of an empty reservoir, a blocked water line, or other failed component. Devices such as the Aquamist DDS2 or DDS3 are recommended here, and for reasons like these it is critical that you purchase your water injection equipment from a retailer with a history of proven water injection technical support.

Water-Injection Nozzle Placement—There are a variety of places that seem to make sense for a water injection nozzle, and you'll hear recommendations on a wide-range of places. The most obvious would be immediately before or after the intercooler, immediately before or after the throttle body, immediately before intake manifold feeds cylinder head, or in parallel with the fuel injection.

The closer you get to the cylinders, the more sensitive the arrangement will be to a variety of factors, so the intake manifold is not ideal. Ditto with anything to do with the fuel-injection setup, not to mention the considerable complexity.

Immediately before the intercooler, while allowing the water to impact the hottest air, is not recommended because of the potential for condensation, the drops of which could obviously affect atomization and cylinder distribution. Immediately after the intercooler removes this potential. A general recommendation made by Aquamist vendors is to place the nozzle right after the intercooler, but as far from the throttle body as possible, allowing the mixture maximum opportunity to affect the intake charge. A criticism of this idea is that while it will have additional opportunity to do exactly that, it also has more opportunity to recondense if allowed to re-cool. This theory would encourage placement closer to the throttle body, ideally in the throttle body inlet, and is generally where most tuners place the nozzle on 3S-GTE applications.

Note: Regardless of whether you place your water-injection nozzle in an intercooler pipe, or the throttle body inlet, be certain to remove the piece before tapping. Even very fine metal shavings can ruin your engine.

Is Tuning Necessary After Water Injection Install?—Ideally, engine management tuning of

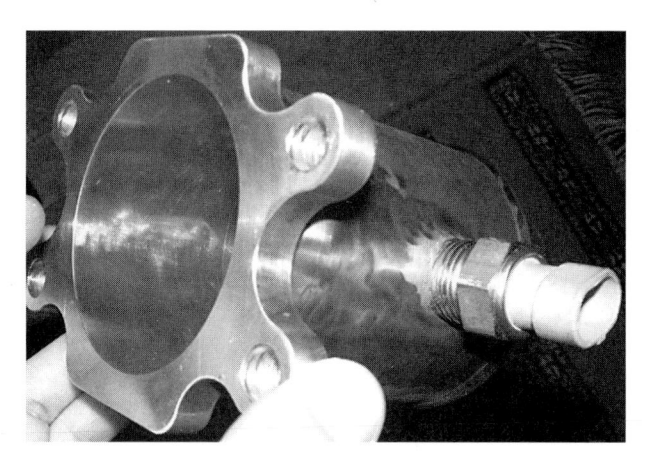

On the 3S-GTE, the throttle body inlet is easy to remove and tap without risk of shavings entering the combustion chamber. It is also an excellent place for water injection nozzle placement due to its location in the intake path. Photo courtesy Jensen Lum.

some kind is necessary after installation of a water-injection system. For best results, water injection must be tuned for. That is, you really need to have some method of controlling the air/fuel mixture and the ignition timing to fully take advantage of water injection's in-cylinder cooling and increased detonation control.

While water injection will enable a factory ECU to allow more timing, more boost and so on, installing one of these systems on a stock forced-induction MR2 will not generally add much in the way of power production. It may add some benefit in that the stock ECU will not have to as aggressively retard timing or limit boost levels, but this depends on a variety of factors: what octane fuel you were running to begin with (California, for example, generally not having access to 93 octane fuel), or the "artificial compression ratio" of your engine, that is, if you had a load of carbon deposits on your pistons tops altering your effective compression ratio, and again, causing the ECU to be more aggressive in retarding timing. Its main benefit on a stock application will be to disallow these instances, and to provide steam cleaning of the combustion chambers, and secondary intercooling. This application of water injection is often called a safety net, where you're not counting on it for performance through tuning, but rather as another method of protecting your MR2's engine.

The most ideal scenario for a water-injection application is on a forced-induction MR2 with a standalone engine management system, a wideband O_2 sensor, appropriate gauges—an EGT and boost gauge for starters—and a dyno-session with an experienced tuner tuning for additional horsepower a properly set-up water-injection system can provide.

Methanol/Water Mixtures—To enhance to detonation-resistance potential of your engine, Toyota installed an intercooler, recommended 91–93 octane fuel, and built-in several features into the ECU and other engine systems to reduce the likelihood—or damage resulting from—detonation. A water injection system injecting pure water is an excellent way to supplement these systems. A water injection system injecting a water-ethanol mix is even better.

Alcohols like methanol have unique stoichiometric properties from gasoline, and simply substituting some percentage of gasoline for methanol will render a richer mixture that must be accounted for to keep from harming power. In fact, the best mixture for performance depends on the charge air temperature, which water injection—whatever mixture you're injecting—directly affects. Because of issues like these, one mixture coupled with one air/fuel ratio cannot be universally recommended, and so, as with many other performance recommendations, testing must be done to find the optimal arrangement for your MR2. We can issue starting points and general rules though.

A 50/50 mixture of water and methanol is a good starting point and for many owners ends up being just right. On a dyno, monitoring EGTs, air/fuel ratios, power production and so on, you can begin experimenting with mixtures if you really want to find the most accurate mixture for your engine. The further you go toward a pure methanol mixture, the higher the octane potential of the mixture, but also the more expensive the mixture.

On paper, one could discuss the theoretical benefits of pure water, or pure methanol injection, but really it boils down to two options: running pure water and leaving some additional power potential out in the name of simpler, slightly less expensive power production, or start experimenting with a 50/50 methanol/water mixture, and continue testing until you're satisfied with your setup based on whatever criteria you set: simplicity and ease-of-use, cost, pure horsepower and so on.

Overall I cannot recommend water injection enough, for any forced-induction car of any level of tune. Even if your MR2 is completely stock, the clean combustion chambers and supplemental intercooling are absolutely faultless benefits that cannot be argued with—unless, of course, you're on the Internet, where everyone argues everything.

Methanol Availability—Methanol is generally available in most hardware stores and other industry-specific retailers, but a shade tree practice common among budget-minded owners is to use

Premixed methanol/water solutions like Boost Juice have some worthwhile advantages over mixing it yourself, namely hassle-free, properly mixed solutions for your water-injection system. Photo courtesy Snow Performance.

windshield washer fluid.

Windshield washer fluids are basically a mix of water and methanol. When this mixture is supplemented with a purer methanol mix like HEET-brand gas-line antifreeze and water-remover, it can work. If you know what you're doing, and don't mind the legwork, this can indeed save you some bucks, but there is some risk in damaging your engine. I would not recommend that you try this.

With the increase in popularity of water injection, many premixed methanol/water solutions are available, such as Boost Juice and Cool Fuel. If you're not into mixing the stuff on your own, and want to be certain what you're using is safe and properly mixed, these are excellent options.

ADVANCED FUEL SYSTEM PERFORMANCE

The consequences of too small or large an exhaust is a change in horsepower; but too little fuel could cost many thousands of dollars in parts and repairs, and has ended many automotive projects prematurely, especially forced-induction projects doubling stock horsepower levels.

Beyond minor tuning of air/fuel ratios through ECU tuning, fuel upgrades are not necessary on all but the most radical normally aspirated MR2s, and so are mainly the domain of the SW20.

Considerable fuel upgrades are necessary for big horsepower, a Stage III SW20 for example, and such upgrades involve a variety of connected components, all crucial for safety as well as performance. The SW20 factory fuel components are stout enough to support significant power increases, but even this flexibility is quickly used when airflow is increased just as quickly through increased boost and larger turbochargers. Items like fuel injectors, fuel pressure regulators, fuel rails, fuel pumps, and a bevy of fuel lines and fittings must be cooperatively upgraded to safely provide radically increased levels of fuel.

Fuel Injectors

The stock Gen II 3S-GTE utilizes 440cc, side-feed injectors, using a resistor pack to allow the use of its low-impedance factory injectors in cooperation with the ECU's high-impedance injector driver. These injectors area able to provide enough fuel to produce just beyond 260 whp. Past that they need to be upgraded to larger injectors, depending on your goals.

With a method of controlling their additional capacity, larger injectors, either the 540cc injectors sourced from the Gen III 3S-GTE, or the 550cc used by the 2ZJ-GTE in the (JZA80) Supra, can be used to increase the amount of fuel available to each cylinder, and with appropriate fuel pressure should be able to fuel 300 whp comfortably, and as much as 320 whp in some setups (but, as we mention below, anything that affects air/fuel ratios is something we need to be cautious about). Past that, 800cc side-feed injectors from SARD, or any number of large, top-feed injectors (see below) are available.

Still, let me be clear here: the only way to know for certain that you have an adequate supply of fuel is to gather data on your MR2 by measuring quantifiables like air/fuel ratio, duty cycle and fuel pressure. Do not depend on time-tested rules when it comes to your fuel or engine management. Utilize the scientific philosophy of observation, theorization, data collection, and further observation and measurement to insure a safe build. A blown engine can force you to sell your car quicker than anything.

And remember, you can't simply insert a new set of larger injectors into your fuel rail—some method of controlling their added capacity must be had. The ECU still thinks it is actuating a 440cc injector, and a 25% increase in capacity will richen an already rich mixture, basically making your SW20 undriveable. This is where the ATS Racing combo kits and ECU ROM-tunes that can control 550cc injectors, or aftermarket engine management systems like the Hydra Nemesis from mrcontrols, come into play.

Note: If you switch to high-impedance injectors, you'll need to bypass the resistor pack. To further confuse the issue, various aftermarket engine management systems can affect what you need. For example, the Hydra Nemesis and Autronic SMC allows owners to run low-impedance injectors without a resistor pack, but others units from other EMS manufacturers may not.

Fuel Rails

The stock fuel rail on the 3S-GTE is a side-feed fuel rail, as opposed to a top-feed rail. In a side-feed setup, the injectors are plumbed into the fuel rail at their "sides." Among the changes Toyota deemed necessary when revising the 3S-GTE were adjustments to the fuel rail, most notably moving to larger bore, though sticking with the side-feed design of the Gen II. This modification by Toyota obviously encourages speculation that the Gen II 3S-GTE's fuel rail might be insufficient to flow the amount of fuel and fuel pressure that 300+ whp demands, considering that Toyota went with a larger bore for only 20 additional hp—and no, the Gen III fuel rail will not fit a Gen II cylinder head.

One way to upgrade the fuel rail is to simply purchase a larger bore side-feed rail from ATS Racing. This increased bore will help support the additional flow fuel pressure potential provided by adjustable fuel pressure regulators, and larger fuel pumps. This will allow the continued use of side-feed Gen II injectors like the stock 440cc, or the 540cc/550cc injectors from the Gen III 3S-GTE/JZA80 Supra, or even available 800cc injectors from SARD as well. If you want a basic rail upgrade, but want to stick with a side-feed design, this is a good choice.

Still, courtesy of Toyota's very methodical overbuilding approach, most tuners have found that the stock fuel rail seems sufficient to supply fuel and fuel pressure to around 275 whp. As we discuss in Chapter 6, at that horsepower point in an SW20, fuel needs overall start changing in a hurry: The bore of the stock fuel rail starts to become inadequate, and must be replaced with a larger bore rail in concert with other crucial changes, including the fuel pressure regulator, fuel pump, fuel injectors, and various fuel lines, rings, and fittings.

At this point, a top-feed fuel rail becomes a better investment. Replacing the factory side-feed injector design, top-feed fuel rails are available, from Wolfkatz Engineering for example. These rails allow for a much broader selection of larger aftermarket fuel injectors, in the same way moving to a larger 17" wheel package opens up the available selection of performance tires. For those owners moving beyond 275 whp, this really is the best path to take.

Note: Full fuel-system solutions from companies

CALCULATING FUEL INJECTOR SIZING REQUIREMENTS

To calculate fuel injector sizing requirements, use the following formula:

fuel injector size = BSFC x hp ÷ duty cycle x no. of injectors

Note: BFSC is brake specific fuel consumption. For example, a 4-cylinder turbocharged engine, producing 300 horsepower, with one injector per cylinder, and an injector duty cycle of 85%, would require approximately:

fuel injector size = (0.60 x 300) ÷ (0.85 x 6) = 35.3 lbs/hr

Remember, if it were easy to build big horsepower engines, everyone would do it. The only way to know for sure if you've got enough fuel is to closely monitor your air/fuel ratio and injector-duty cycle.

Formula courtesy RC Engineering

---

like Wolfkatz Engineering relieve possibly the largest headache when piecing together your own fuel system: fuel lines, O-rings, and fittings. The stock fuel return line may not be compatible with many aftermarket rails, specific O-rings, clips, wiring, and other fittings can not only add up in sourcing, purchasing, and shipping cost and time, but in compromised quality as well.

Dual-Feed Fuel Rail—One theory common in the MR2 tuning community years ago to help explain a disturbing amount of blown 3S-GTEs involved the dual-feed fuel rail, or DFFR. It was suspected that the Gen II 3S-GTE fuel rail starved the #3 and #4 cylinders due to the sequential fueling pattern of the stock arrangement. Being fuel starved, they were thus "lean," and a lean air fuel ratio means heat. Combined with the OEM intake manifold design that flows somewhat unevenly across the four cylinders as well, it was theorized that a dual-feed fuel rail—one that feeds the rail at both ends—would help correct this lean issue, and engines would stop going boom.

The dual-feed fuel rail did its job commendably, but did not stop 3S-GTEs from "popping," so the trouble—were it in but one spot—had to be elsewhere. Today, a dual-feed design is still useful when feeding a large amount of fuel through large injectors, but is no longer considered a savior against a grenading engine.

Fuel Pump, Fuel-Pressure Regulator

Sufficient fuel pressure is necessary to maintain not only a proper air/fuel mixture, but proper injector atomization as well. Visualize the dribble of a hose with poor pressure, versus the strong spray pattern of one with appropriate pressure. This type of pressure in the SW20 fuel system occurs through a coordination of the fuel system components, the fuel pump being the workhorse of this system. Upgrading the fuel pump can become necessary as the need for fuel increases with increased airflow—boost.

There are a couple of popular SW20 fuel pump upgrades, namely the Walbro 255 lph, and the OEM fuel pump from the big-brother JZA80 Supra we also took the injectors from. Still, the stock SW20 fuel pump is actually fairly stout, not unlike the rest of the car. How much horsepower it will support actually depends on several tangent factors, most having to do with fuel pressure. Like traditional fluid dynamics, pressure directly affects flow. The SW20 has a base fuel pressure between 40–42 psi, set by the stock fuel pressure regulator. The fuel pressure regulator does exactly what it sounds like it would—regulates fuel pressure for safe combustion-event fueling.

Fuel pressure, measured in the rail and at the fuel injectors themselves, is figured by adding base fuel pressure, and the manifold pressure. Manifold pressure on a forced-induction car is obviously a factor of boost pressure, as the supercharger or turbocharger is pressurizing the intake manifold with air at some pressure beyond standard atmosphere, so the formula looks something like this:

base fuel pressure + manifold (boost) pressure = fuel pressure

Unlike an aftermarket rising-rate fuel pressure regulator, the stock fuel pressure regulator has a 1:1 ratio. This means that for each psi of manifold pressure, fuel pressure is increased by that amount. A car with stock fuel pressure, let's use 42 psi, running 18 psi of boost pressure, would then be operating at a fuel pressure of 60 psi.

An adjustable fuel-pressure regulator is optional to increase the fuel pressure on smaller injector setups to increase their "life." For example, 550cc injectors that were maxed out around 300 whp may, in the short term, be stretched to fuel another 20–30 whp with an increase in fuel pressure. This, however, is not the best approach. Ideally, correctly sized fuel injectors, operating at 85% duty cycle or less, will be supplied fuel by a pump and rail able to sustain adequate fuel pressure at any boost level your chosen turbocharger is capable of realistically operating at. In simpler terms, your turbocharger system should not be more powerful than your fuel system that sustains it.

Note: If you've removed or replaced anything in your fuel setup, it's always a good idea to check for leaks without the heat and spark of a running engine. To do so, insert a paper clip in the diagnostic ports between +B and FP and turning the key to the "ON" position. This will charge the pump and pressurize the system.

ADVANCED CAMSHAFT & VALVETRAIN PERFORMANCE
Camshafts
The Idea: Improved cylinder head breathing

Engines are very simply glorified air pumps—complex, heavy, inefficient air pumps at that. The internal contents of engine block and cylinder head determines the displacement of the engine—that is, the total volume of the cylinders. Since the engine is an air pump, the amount of air that the engine can process is determined by a function of that displacement against rpm—how many times per minute the engine pumps that total volume of air in and out of the cylinder head.

If the engine block is the lungs, the cylinder head can basically be thought of as the nostrils. All MR2 engines are dual overhead cam setups, which means two camshafts are located in the cylinder head, as opposed to a single camshaft within the engine block in the case of the Chevy small-block, for example. The cylinder head is responsible for ushering the new air in through the air filter and intake manifold, and the old, spent air out past the exhaust valves, out the exhaust manifold, and eventually out the exhaust system. The critical performance role of the head is easy to understand.

HKS makes aftermarket camshafts for the 4A-GE, as do Toda, Web, Crow and a number of other manufacturers. The main differences are in the duration and lift. Photo courtesy HKS USA.

4A-GE

As we've discussed, the 4A-GE series engine by Toyota is a free-revving, energetic powerplant that was way ahead of its time when Toyota dropped it behind the passenger compartment of the MR2. Keep in mind that in 1985 Honda, the most vivid celebrator of technology in today's automotive circles, still snuck carburetors under the hood of various product, including its then-flagship Prelude. Factoring in cost and application, the 4A-GE was nothing less than a marvel. Displacing less than a 2-liter of Cherry Coke, and plain ol' normally aspirated in 4A-GE guise, its best characteristics are its simple, compact design and its willingness to rev. This engine was a blessing to the industry, and helped usher in an attention to high-revving, small displacement, DOHC design that continues to evolve today.

Still, the 1980s are long gone, and almost every vehicle Kia produced for the U.S. market in 2009 makes more horsepower than the 4A-GE. If you want brute power—and it has to be in an AW15—your best option is to swap in a completely different engine.

The stock cams are an excellent compromise. When designing a suspension setup in a production automobile, suspension engineers must consider a wide variety of factors, chiefly performance, ride quality and cost. Powerplant engineers must also strike a similarly delicate balance of emissions, idle, N.V.H., low-end torque, midrange surge, and upper rpm hp when designing an engine and camshafts as well. And we're not even getting into the far more complex technology of the VVT-i engines. So when do you mess with the magic camshaft formula Toyota worked so hard to develop, and change out camshafts?

The most common scenario is this: we have a healthy AW15, with a basically stock, healthy engine. We are looking to exact enhanced character and performance (character is purposefully listed first) from our little sports car. In this case, we obviously don't require the same compromise Toyota engineers had forced on them when designing a car for both track junkies and commuting daily drivers. The gains from camshaft upgrades on the 4A-GE are relative to your expectations, previous experiences with similar upgrades in other cars, bang-for-buck and the meager stock hp output the 4A-GE to begin with.

On a stock 4A-GE, you can do as badly as lose torque and horsepower over your entire powerband, or you might gain as much as 6–10 peak whp. You can usually expect between 8 and 20 whp from a camshaft upgrade on a slightly modified 4A-GE,

Camshafts are a worthy upgrade on a highly modified engine. They basically improve how much air the cylinder head can process, and will generate horsepower if there is need for the extra airflow. 3S-GTE with HKS camshafts is shown.

depending on the rpm you've built the engine for, and other associated modifications.

Note: Do not choose camshafts based on brand, but rather by their lift and duration, matched to the valvetrain and bottom end you are going to match them with, and the powerband you are seeking. It does you no good to buy super-trick camshafts that make power at 8000 rpm if you are not installing them in an engine built to support that rpm level.

3S-GTE

For most mild to moderate forced-induction builds, camshaft upgrades are not necessary. A forced-induction engine has far more flexibility in its tuning. There is more fertile ground elsewhere—namely the turbocharger—which makes a camshaft upgrade, a fairly complex upgrade, less critical. Think cost-benefit analysis, and be sure to factor in costs beyond simply the purchase price of the camshafts themselves: think valvetrain, cam gears, labor, tuning, etc.

It is less complex and more cost-effective to upgrade the turbocharger itself, and then to focus on engine management tuning than it is to start experimenting with various camshaft grinds, and all of their associated installation issues such as degreed installs, cam gear settings, staggered setups and so on.

Turbocharged applications like the 3S-GTE must deal with much higher exhaust backpressures than the normally aspirated AW15 and SW21. The turbocharger is a wonderfully helpful device, but its placement in the exhaust stream somewhat alters the physics of the air's behavior in the cylinder head, backing things up, involving complex issues like reversion, valve overlap, and other factors fully

Toyota MR2 Performance

Cam gears basically allow you to advance or retard your camshafts to tune the timing. These are HKS cam gears adjustable markings, in degrees.

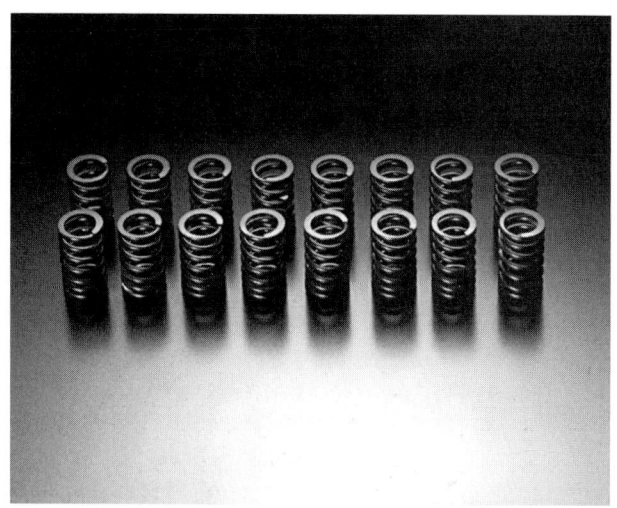

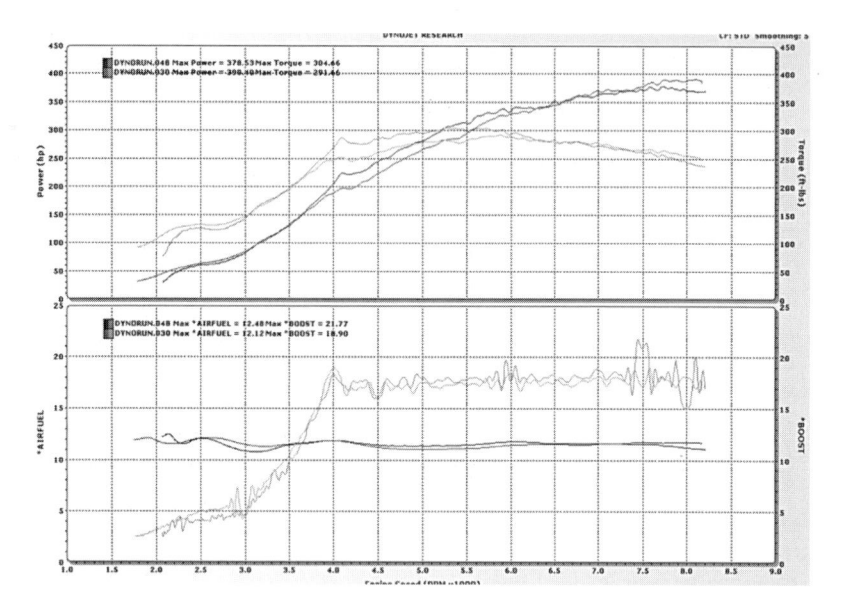

As a result of cam gear tuning, this 3S-GTE picked up a consistent 12 lbs-ft. through significant portions of the powerband. Drawing courtesy ATS Racing.

Valve springs, such as these springs manufactured by Engle, are a necessity if you want to upgrade camshafts to anything other than mildly increased lift. Choose the right set for your budget—many sets require modifications elsewhere in the cylinder head. On 3S-GTE applications, these Engle springs do not. Photo courtesy Engle.

understood by few non-mechanical engineers. This complexity, and the concert of overall engine setup and design, must be respected. For now, just consider that turbocharged engines can thrive with much milder camshaft designs than the radical camshafts high-performance normally aspirated engines must employ.

Note: Adjustable cam gears are critical to extracting optimal performance from your aftermarket camshafts. In fact they can even be used on stock camshafts for performance gains as well, allowing to to advance or retard the camshaft timing several degrees in either direction.

Note: The cam gear 3S-GTE dyno chart also demonstrates the benefit of "area under the curve"

over peak hp, with the "after" run down around 11 peak hp. When looking at the swipe of the tachometer throughout the rev range, the vast majority of powerband is improved; the peak hp, while down, only represents a very narrow band just prior to redline. For reference, the numbers on that dyno are as follows: before cam gear install and tune, 390 hp, 292 lbs-ft.; 379 hp, 305 lbs-ft. after.

Valvetrain

The Idea: Controlling the movement of the intake and exhaust valves

Any MR2 cylinder head is a busy place with 16 valves and valve springs, retainers, shims, buckets and guides all hammering away. Higher-lift cams with longer durations often require aftermarket springs designed for those specifications. Larger turbochargers require lots of air to motivate, and those that don't fully spool until beyond 5000 rpm require higher rpm that the stock valve springs can withstand. Rev it high enough, generally just past 7500 rpm, and the stock 3S-GTE valves will begin to "float," never fully closing before having to open again. At 7000 rpm, each valve must open and close 58 times per second!

When you factor incredible combustion chamber pressures from a bunch of boost, stiffer valve springs are a must if you're running a bigger turbo, high lift/long duration cams, and a lot of revs. As a rule of thumb, if you upgrade your camshafts, upgrade the valvetrain. In fact, many cams,

3S-GTE CAMSHAFT SPECIFICATIONS

The camshaft's job is to actuate the valves, directly so on an overhead cam setup like all MR2s. What makes one camshaft different from another? If you're cam-shopping for your 3S-GTE, consider the following specifications for available camshafts. For reference, consider the following camshaft specifications offered by MR2-tuner Bryan Moore.

Stock Gen II 3S-GTE (Intake & Exhaust)
Advertised duration 236 degrees
Duration @ 0.010" lift: 260.0 degrees
Duration @ 0.020" lift: 211.0 degrees
Duration @ 0.050" lift: 160.5 degrees
Max Lift: 8.52mm (0.335")

Stock Gen III Intake
Advertised duration 240 degrees
4 degrees more duration, .5mm more lift than the Gen II

Stock Gen III Exhaust
Advertised duration 236 degrees
Duration @ 0.010" lift: 260.0 degrees
Duration @ 0.020" lift: 211.0 degrees
Duration @ 0.050" lift: 160.5 degrees
Max Lift: 8.52mm (0.335")

HKS 256 Cams
Duration @ 1mm lift: 216 degrees
Max Lift: 9.0

HKS 264 Cams
Duration @ 1mm lift: 224 degrees
Max Lift: 9.2mm

HKS 272 Cams
Duration @ 1mm lift: 232 degrees
Max Lift: 9.4mm

HKS High Lift 256 Cams
Duration @ 1mm lift: 216 degrees
Max Lift: 9.5mm intake, 9.2mm exhaust

HKS High Lift 264 Cams
Duration @ 1mm lift: 224 degrees
Max Lift: 10.0mm intake, 9.7mm exhaust

HKS High LIft 272 Cams
Duration @ 1mm lift: 232 degrees
Max Lift: 10.4mm intake, 10.2mm exhaust
The HKS Gen III 272 "high lift" cams use a lobe centerline of 111 deg ATDC intake, and 116 deg BTDC exhaust.

JUN Cams
Part # 1004M-T001 (intake) 1004M-T101 (exhaust):
256°F duration 9.0mm lift
Part # 1004M-T002 (intake) 1004M-T102 (exhaust):
264°F duration 9.0mm lift
Part # 1004M-T003 (intake) 1004M-T103 (exhaust):
272°F duration 9.3mm lift

---

depending on their specifications, mechanically require it.

The valves basically inhale and exhale for the engine—two intake valves per cylinder opening to allow the air/fuel mixture in on one stroke, and two exhaust valves per cylinder opening on the exhaust stroke to allow the spent combustion gases to be expelled.

For stock configurations this works well—in fact, the 4A-GE, 4A-GZE, and 3S-GTE all feature exceptional cylinder heads that aren't significant restrictions until stock horsepower levels are well surpassed. Once you've radically increased the amount of air your engine is being asked to process though, whether through rpms, aggressive camshafts, additional boost, or all three, the size of the stock valves can become a restriction. In these cases, aftermarket companies like Ferrea manufacture over-sized valves that are a bit larger—usually 1mm oversized—to enable increased airflow processing. Cost-benefit analyses in most of my builds have suggested that I leave my cylinder heads mostly stock—valves, valve springs, and so on. As turbocharger sized increases beyond "+1" upgrades like the CT20B, and even TD06 and T3/T4 turbochargers, increased rpm become more of a necessity, at which point enhancements to the cylinder become more useful.

ADVANCED TURBOCHARGER PERFORMANCE
Turbochargers
The Idea: To Increase power by increasing the air charge available for combustion
SW20—In the staged buildup for the SW20 we discuss a variety of turbocharger considerations, and

The CT20B and CT26 both share the inventive twin-scroll design, which minimizes convergence of exhaust pulses, reducing turbulence and increasing spool. Photo courtesy Brian Hill.

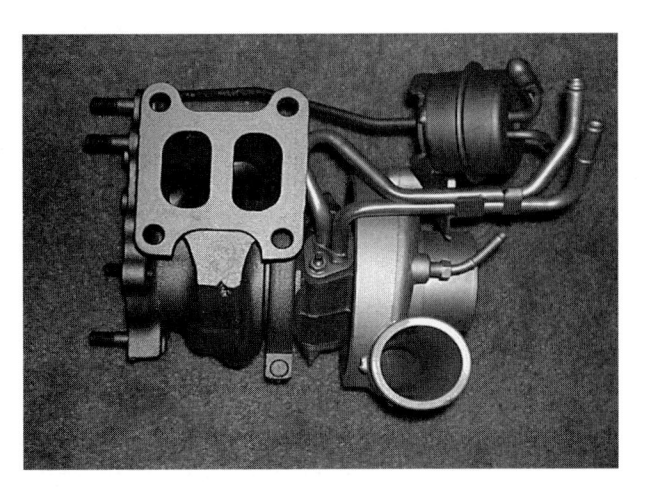

The easiest way to differentiate the CT20B turbocharger (pictured) from the CT26 is by the number of downpipe bolts: the CT26 has six, the CT20B seven. Nonetheless, CT20B and CT26 downpipes are interchangeable—it is simply a matter of removing a bolt. Photo courtesy Brian Hill.

take a look at a variety of kits that can help you achieve any of your power goals. Some may genuinely wonder, though: if you own an SW20, of course you already have a turbocharger that provides boost. What's the matter with the turbo you've got? The answer lies in that turbo's efficiency.

The SW20 features an internally gated CT26 turbocharger whose range of dynamic performance is insufficient even for stock boost levels. While the twin scroll technology is useful in separating exhaust pulses to smooth exhaust flow and assist the turbo in spooling, basically reducing the convergence of exhaust in half, at stock boost the CT26 falls on its face past 5500 rpm or so, rendering a very Pontiac Grand Am V-6-like powerband.

The CT26's main shortfall lies in its relatively tiny exhaust housing. Though the generously sized compressor is able to build much higher levels of boost, the exhaust housing simply cannot process enough air to keep from choking the engine at elevated rpms. Turning up the boost will indeed increase horsepower, but will generally have a more significant impact on the lower half of the powerband, increasing torque substantially, and—just like stock—falling off painfully above 5500 rpm.

This is what is known as a bottleneck, and for years owners tried to fix this malady by replacing the exhaust housing and wheel with larger pieces, which makes sense. But there is more to the story here than bigger is better, and these upgrades typically offered very minimal gains at costs they did not justify. In fact, it wasn't until 15 years after the SW20 was produced that the aftermarket has been able to produce an upgrade to the CT26 worth owning, ATS Racing's aptly named CT27.

Toyota must've sensed a similar inadequacy in the CT26, because it was among the most significant changes on the Gen III 3S-GTE, replaced with the CT20B that, while still stock and relatively modest, offered substantial increases in airflow and performance. The CT20B can move enough air to allow the 2.0L 3S-GTE to pull hard all the way to redline, while still spooling as quickly—or quicker—than the CT26. This is accomplished through the proper matching between compressor and exhaust housings, and use of advanced materials on the turbocharger wheels. The CT20B comes in two variations: one with a steel wheel, and one with a ceramic wheel. Ceramics were chosen for their light weight and superior thermal dynamics, but over the lifespan of the Gen III, this was also shown to be a detriment: at elevated boost levels, the ceramic wheel themselves have been known to structurally fail, obviously ruining the turbocharger, and likely the engine as well. Though many tuners have run ceramic-wheel CT20Bs at 18+ psi for years without issue, the steel-wheeled CT20B is favored by the performance-minded SW20 owner for durability's sake.

AW11/ZZW30—For AW11 owners—supercharged or normally aspirated—there currently, as of this book's writing, is no complete, bolt-on aftermarket turbo kit, and at this point in its life span, there may well never be. There are many industrious owners out there who spend their weekends in boneyards pulling turbochargers and intercoolers from old Saabs to make "custom" turbo kits. And there are those who go the higher-dollar route, but essentially do the same thing, source their kit in pieces from various manufacturers, like ATS Racing, Speed Source, KO Racing and AutoLab for example. The problem with the AW11 (besides rust), is the math: the number of AW11s

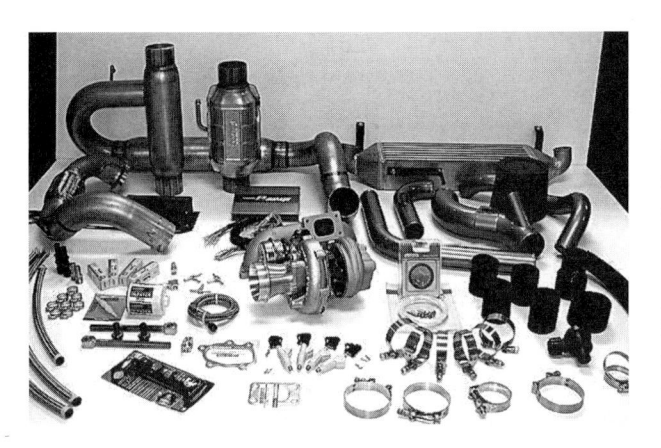

Advanced MR2 Performance Modifications & Theory

When each individual part of this ZZW30 Hass Pro Turbo kit is viewed, the amount of engineering and development in a properly thought-out turbo kit is impressive. Photo courtesy Hass.

on the road is, at best, falling slowly, making the market for its products a dead-end for most manufacturers, especially as compared to the market potential of more current and readily available Hondas, Scions, Nissans, etc.

And thus Hass Pro Turbo's attention to the MR2 Spyder with their very thoroughly developed kit, and not the classic of classic AW11. As with all performance parts, from exhaust systems to clutch setups, turbocharger kits sometimes come into and out of production quickly.

Turbocharger Efficiency—A larger, more efficient turbocharger will allow you to raise the inlet pressure while lowering the outlet pressure—that is, outlet temperature. This allows one turbo to make significantly more horsepower at the same boost level as another, and even a modest turbo upgrade like a CT27 or CT20B is a fantastic upgrade even for the most satisfied SW20 owner. Larger turbochargers can also hold a set boost pressure all the way to redline, offering up a flatter torque curve, rather than one that peaks early and falls off.

Just as camshaft alterations can affect that shape of a curve on a normally aspirated engine, turbocharger changes directly manufacture the curve on a turbocharged car. With most well thought-out and developed sports cars—especially those driven mostly on the street—one should strongly consider curve every bit as closely as peak horsepower.

Horsepower Curve vs. Peak—The curve is basically the shape of the powerband. Peaky engines are described this way because while they may offer impressive horsepower on paper—394 hp, for example—this is only a peak reading. A peaky curve only provides impressive acceleration for a small fraction of the powerband, and this is a discussion relevant not only to turbochargers, but for camshafts as well. This yields little area under the curve—or a swathed portion of the powerband where the turbocharger has left its mark.

A flat torque performance by the engine is preferable to some for this reason; it makes even power from low to high rpm, giving you ample power throughout the range, sacrificing the height of that peak for the accessibility of a broad powerband. For a street-driven MR2, this is a reality to consider, and is a perfect example of the subjective nature of tuning.

There is also the "just downshift" philosophy: when you want big power, use your shift lever to find a gear that will put your turbocharger in its sweet spot. Simple enough for some folks who realize that a peaky powerband is a physics-based reality of a small-displacement engine, and that it doesn't have to limit the performance of your car the way it may seem to on paper. After all, what is faster on a road course—an MR2 with the stock CT26 that makes 223 lbs-ft. from 2600–5200 rpm, or a 20G-powered MR2 that makes at least 285 lbs-ft. from 3800 to 7000 rpm? In this case, the only downside to the larger 20G turbocharger is what happens at 3500 rpm and below, to which you may respond "who cares?"

The Street Brawler from KO Racing is an excellent mid-range turbo kit for the MR2, capable of comfortably producing anywhere from 250–350 hp. Note the adapter that enables use of an external wastegate with the stock exhaust manifold, and the flexpipe on the wastegate dump tube. The devil is in the details. Photo courtesy KO Racing.

Wastegates
The Idea: Precise Boost Control
The function of a "wastegate" is simple enough to guess at. It is a gate that allows boost "waste" to pass. There are internal and external wastegates.

A wastegate controls how much boost the turbocharger generates. Without a wastegate, as exhaust gases encountered the turbine, boost would simply continue to build, the added horsepower from smaller initial amounts of boost providing further exhaust gas energy to spool the turbo to even greater heights and more boost, and so on.

Stock turbochargers—and many aftermarket ones like the old GReddy TD06SH kit— feature internal wastegates. This means that the wastegate is located within the turbocharger assembly. The benefit of an internal wastegate is in their simplicity and cost-effectiveness. They require far less plumbing and fewer parts than an external wastegate setup, making them a very nice choice for your average Stage I or Stage II MR2. Both the stock Toyota CT26 and CT20B feature an internal wastegate, as do most stock turbochargers.

External Wastegate Basics—Then why aren't all turbochargers internally gated? Internal wastegates are great for lower boost pressures and the mediocre airflow capabilities offered by smaller turbochargers, but when the airflow increases courtesy of a lot of boost, or more dramatically, a lot of boost from a large, efficient turbocharger, internal wastegates struggle to keep up with the airflow demands. Their "gates" simply aren't big enough. In cases like this, an external wastegate is necessary.

An external wastegate is independent of the turbocharger, and is large enough to bleed off larger quantities of air. How exactly the wastegate works somewhat depends on the exact setup. External wastegates are manufactured with a certain tension spring designed to open once a certain boost level has been reached. This is a mechanical operation, and is non-adjustable without physically altering or replacing the wastegate. Once the boost level the wastegate was designed to open at is reached, any other boost the turbocharger builds is simply bled off back into the exhaust stream, and conveniently returns to the atmosphere.

External wastegates can be either vented to the atmosphere, or vented back into the exhaust system. The former is very loud, but is simpler to plumb, and offers an aural warning with far more credibility than a blow-off valve. Routing it back into the exhaust stream will almost entirely mute the sound of a wastegate's operation. Of the available external wastegate selection, TiAL is the industry standard.

This external wastegate manufactured by TiAL demonstrates the substantial size difference compared to internal wastegates. Their ability to process large amounts of air is easy to visualize. Photo courtesy TiAL.

External Wastegate Adapters—While many top-end aftermarket turbo kits provide aftermarket exhaust manifolds with custom methods of plumbing an external wastegate, an easy way for most 3S-GTE owners to install an external wastegate is to use an adapter to place the wastegate. This adapter is plumbed in close proximity to the exhaust manifold where all four runners converge, providing sufficient exhaust flow to ensure precise boost control, basically sitting between the manifold and turbocharger itself, providing the wastegate with a place to be anchored, and access to all of the exhaust flow it requires to control boost levels.

The critical aspect of these adapters is fitment. Sitting between the turbocharger and exhaust manifold, it physically displaces the turbocharger somewhat, how far and in what direction depending on the dimensions of the adapter. If the adapter's dimensions aren't just right, it is very possible that it could place your turbocharger in contact with clutch slave cylinder lines, driver cabin firewall, or other things your turbocharger isn't supposed to be touching. Verify fitment before purchasing any adapter!

TiAL External Wastegates—External wastegates are roughly separated by their flow capabilities, a 46mm external wastegate for example able to flow more air than a 40mm unit. As a rule of thumb, the larger the turbocharger, the larger the wastegate needs to be.

TiAL manufactures 38mm, 40mm, 44mm, and 46mm external wastegates, with unique springs available to render a variety of boost pressures from 3.6 psi to more than 23 psi. The downsides to external wastegates are cost and complexity. Not only do they add considerable cost to your turbocharger system, with parts and plumbing as

ADVANCED MR2 PERFORMANCE MODIFICATIONS & THEORY

much as half or more than what a quality turbocharger cost—and nearly as much as a CT20B costs itself—they also require use of either an aftermarket exhaust manifold, an adapter for the stock manifold, or modification (welding) to the stock manifold.

Note: Regardless of spring tension or claimed boost pressure the wastegate allows, boost pressure should always be monitored with an aftermarket boost gauge.

ADVANCED ENGINE MANAGEMENT PERFORMANCE
Standalone Engine Management Systems

The Idea: More intelligently and efficiently controlling an increasingly complicated combustion process

The factory ECU, located in the trunk under the trunk carpeting, if you've ever wondered, is fine when controlling factory parts, but when you are asking your stock ECU to coordinate an effort between aftermarket camshafts, cam gears, turbocharger, intake manifold, fuel injectors, exhaust, air filter, throttle body, and intercooler, it loses its ability to ideally handle that workload.

With this in mind, rather than the stock ECU burdened with a half-dozen black boxes, an entirely new engine management system is often the best choice. These systems are much more at home dealing with all of the extras highly modified systems employ, including additional features like water injection, boost, launch and traction control. Their most important feature, though, lies in their ability to design completely custom engine management maps. Differences across these systems are noted in greater detail in the staged buildups, but for now be clear about the sheer variety of bells and whistles they offer.

In addition to (spark) timing, engine management is responsible for dictating an air/fuel ratio. The higher the ratio, the leaner the mixture, the hotter the exhaust gas temperature, the more dangerous the tune, while richer mixtures run cooler and cost power. In fact, the fuel portion of the tuning process can be reduced, in part, to a process where the mixture is incrementally leaned (to a point, anyway) to make additional power.

Engine management maps are basically numerical values in a giant grid—values that dictate when the spark plugs fire, when the injectors fire, and so on. This allows each engine to have its own custom-charted computer, rather than a one-size-fits-all ECU Toyota gave it. In this way, custom air/fuel ratios and timing values can be designated, rendering not just increased performance, but long-

Generally speaking, the larger your turbocharger, the larger your external wastegate should be so that you can properly hold and maintain a chosen boost level. External wastegates come in a range of sizes to suit a variety of needs. Photo courtesy TiAL.

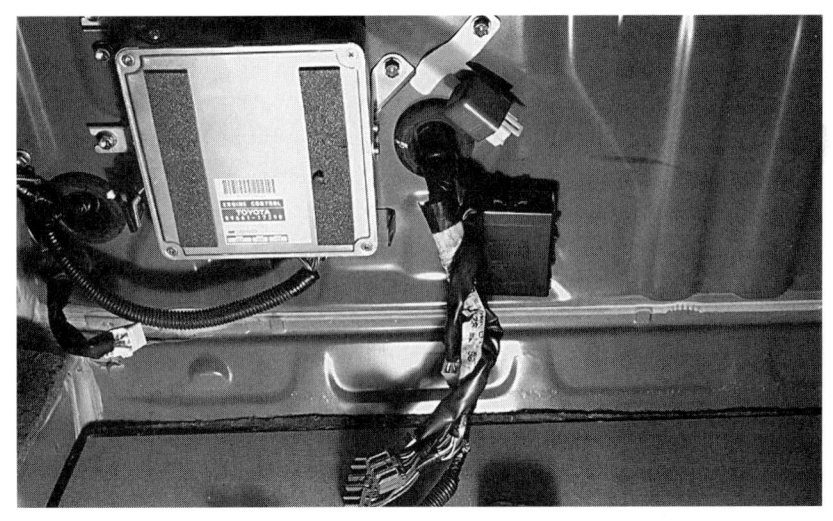

The factory ECU, both directly and indirectly, is responsible for orchestrating the combustion events in your MR2, and besides some doing some funny things with timing on the Gen II 3S-GTE, does a fine job stock. Once you've changed out all the parts the ECU was designed to have control over, it makes sense to consider replacing the ECU as well.

term durability as well.

Like a coil-over suspension, though, having all that adjustability won't help—and could very well hurt—if you don't know what you're doing. Tuning an engine management system isn't as simple as experimenting with shock settings, tire pressures, or alignment, or as free of significant consequence. An EMS is only as good as the person tuning it.

Engine Management Technical Support—In response, vendors and knowledgeable grassroots-level people that know their stuff as well or better than traditional vendors have developed harnesses and plug parts that allow many systems to be "plug

Toyota MR2 Performance

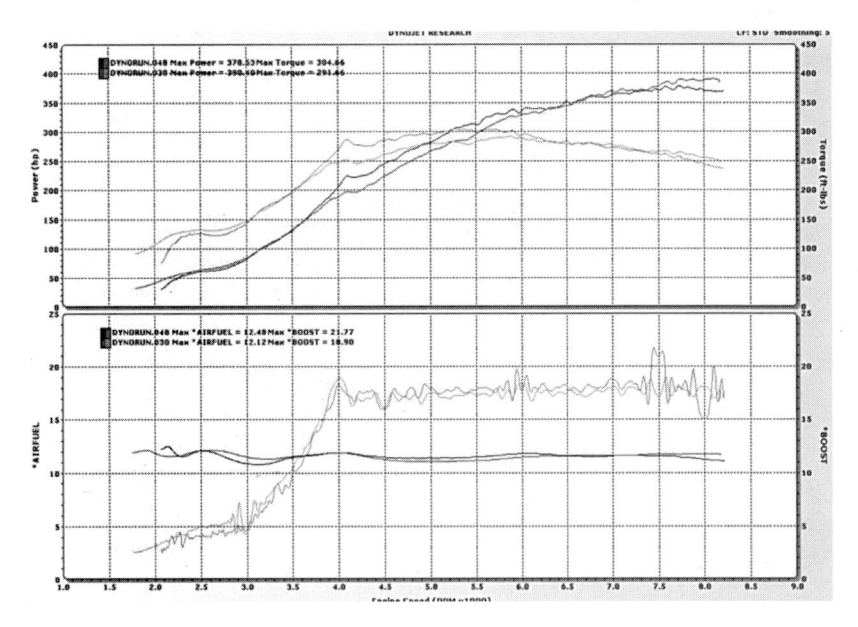

This dyno test of a very potent 2.2l stroker 3S-GTE setup owned by Mark Sink and tuned by Bryan Moore shows that on this run, the ratio never went above 12.48:1 (the lowest of the bottom portion of the graph). Should non-optimal ratios develop, with a standalone engine management system, an adjustment is only a few keystrokes away. On this run, the 3S-GTE produced 379 hp, 305 lbs-ft. at 22 psi. Image courtesy Mark Sink/Bryan Moore.

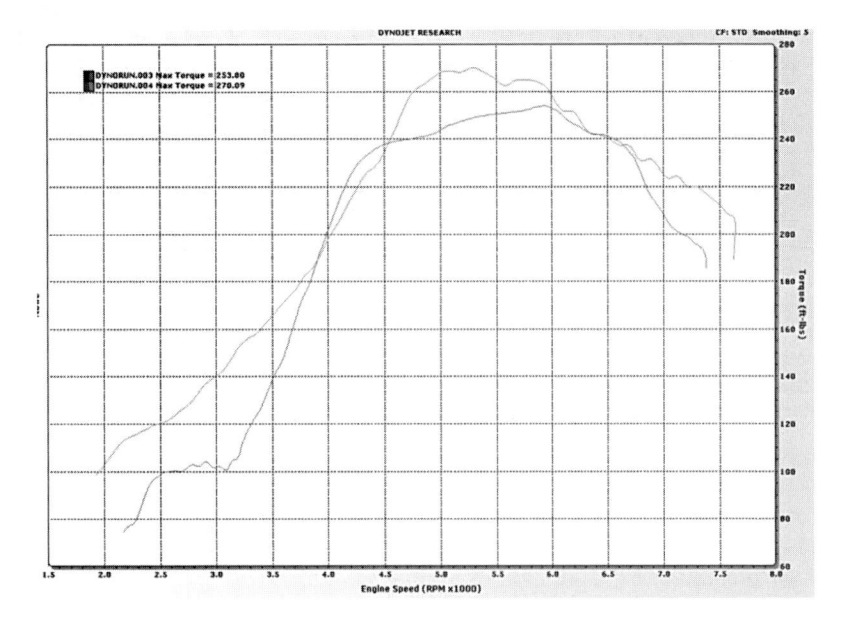

This stock ECU versus standalone engine management system dyno demonstrates the effect a well-tuned engine management system can have on the torque curve of on even a mildly modified MR2, 254 lbs-ft. before, 270 lbs-ft. after. Both turbocharger spooling and peak horsepower were significantly improved. Image courtesy Bryan Moore.

Aftermarket engine management systems are complex devices whose install and tuning are best left to experts—especially those like the Electromotive Tec 3 that also alters how spark is delivered.

and play"—installation amounting to the mounting of hardware, some very limited wiring and a simple "plugging in."

They also ship with base maps, or maps tuned enough for the car to start and roughly run, but not optimized for your application, something the progression of the Internet has helped with tremendously.

A major factor to consider when choosing an EMS system is technical support. The more common a choice an EMS is—on MR2s of your generation, for example—the more support you'll find. Though technical support from vendors or manufacturers is critical, the most important kinds of support in regards to EMS tuning lie with tuners and other enthusiasts. Having accessible tuners who know the system you are considering is important. While many systems are similar, they are enough differences between them that it is important that a tuner not only know the EMS, but the engine he is working on as well. If you find a reputable tuner of EMSs on LS1 V-8s or 4G63s from Mitsubishi, that is better than a tuner with no experience at all, but not as ideal as finding a tuner that knows your MR2 engine and its character.

For example, tuners like Stephen Gunter, John Reed, Bryan Moore, Ricky Benitez, and Aaron Bunch have been working on the 3S-GTE for years, and this knowledge should not be discounted in its usefulness and your buildup. Nor should the popularity of an EMS in your respective MR2 community. Choosing an engine management system is not the time to set trends unless you have a very good reason for doing so. When you run into

WIDEBAND O₂ SENSORS

Most modern combustion engines—including those in your MR2—have OEM O₂ sensors that roughly evaluate the exhaust gas, and adjust the air/fuel ratio of the engine accordingly. These are also called lambda sensors. This measurement is done very crudely, though, via a narrowband O₂ sensor. These ceramic-coated sensors roughly report basic air/fuel mixture back to the ECU, which then uses this information to help adjust the air/fuel mixture, mainly for emissions, gas mileage, and idle considerations.

A wideband O₂ basically enhances the communication to whatever engine management you are running. A wideband sensor allows for extremely precise measuring of the exhaust gas, offering accurate air/fuel ratio data for engine management tuning. Wideband O₂ sensors used to be incredibly expensive, but have dropped drastically in price in the last several years. If you've got the means to adjust your air/fuel mixture, timing, and boost based upon the information it provides, it is an excellent tool to buy that can pay for itself very quickly. A greatly reduced need to for a dyno and their wideband O₂, a wideband O₂ and accompanying data-logging (also available as an option on such systems for very reasonable prices) are must-haves if you have engine-management capabilities.

For dyno-tuning, a wideband O₂ sensor is a must. Though these have traditionally been supplied by dyno-facilities, they have become increasingly affordable over the years, now available for less than the price of an aftermarket exhaust system.

Most modern production engines feature an O₂ sensor, many more than one, to monitor and adjust air/fuel ratios. Factory MR2 O₂ sensors operate over a very narrow range however, offering crude information to the ECU. A wideband O₂ sensor reacts much more quickly with far more resolution, critical for safe, precise tuning.

trouble with your EMS—and you will—having other folks available with the same EMS on the same engine in the same car is a priceless resource.

Note: Wideband O₂ sensors have become much more affordable in recent years, thanks in part to companies like Innovate Motorsports.

ENGINE SWAPS

The most certain way to significantly upgrade the performance of your AW15 or SW21 is an engine swap, and with the growth of the MR2 community, and the development of plug-and-play wiring harnesses from companies like mrcontrols and Phoenix Tuning making the wiring of such swaps less of a headache, engine swaps are not as daunting as they once were.

As previously discussed I generally hesitate to suggest camshaft upgrades in the AW15 in all but the most serious setups. While certainly capable of producing horsepower, the somewhat limited bang-for-the-buck offering of a set of cams pales in comparison to the promise of what an engine swap can deliver. Granted engine swaps are very labor-intensive, whether you're turning the wrenches yourself or paying to have it done, so they are not for everybody.

Reliability concerns and hacked installs can make the idea seem downright nightmarish—swaps can and do go bad. In the end, though, if a significant increase in power in your AW15 is an absolute must, a dead-stock 20-valve produces the same amount of power as a very built 16-valve engine—and dead-stock, contrary to the nomenclature, generally means reliable. Pushing the stock engine to within an inch of its life, out of pride or stubbornness, doesn't make as much sense as dropping in an engine designed to easily produce the horsepower you are after. Again, tuning

Toyota MR2 Performance

> ## DATA-LOGGING
>
> Data-logging is the important ability to retain a record of the engine's performance—not horsepower, but air/fuel ratios, detonation events, timing and so on. After a dyno run, for example, a laptop can be used to download this record for analysis and future tuning by a professional.
>
> The APEXi Power FC is a plug-and-play standalone EMS with handheld interface called the "Commander." Unlike most standalone EMS's, there is no data-logging with the Power FC out of the box—that is, no retaining of data for the use of future tuning adjustments. The APEXi Power FC is designed for ease-of-use, and data-logging was obviously seen as non-crucial by APEXi.
>
> Luckily, this can be remedied through the use of a system developed in New Zealand called the Datalogit that allows the Power FC to interface with a laptop for tuning and data-logging. With Datalogit, the Power FC becomes an easy-to-use, powerful engine management system.

Clips can also come with other goodies beyond engines, such as these revised SW2x taillights and aftermarket exhaust. Most importers have gotten wise to this, however, and remove these parts for independent sale.

Beyond your local boneyard, engines and other parts necessary for an engine swap are also available from importers that import clip. A clip is a portion of the car, usually containing the engine, that is literally cut away from the rest of the car for easier shipping. On the MR2, clips are obviously "rear clips," while those housing 20-valve 4A-GEs are "front clips," the 20-valve produced in front-drive configurations from the factory.

philosophy applies here. Ultimately, you should always build your MR2 exactly how you want it—just be aware you very well could be jacking everything up in the process.

This certainly applies to the SW21 as well. 5S-FE headers, cams, and other upgrades may satisfy a small percentage of owners with very modest power increases in mind, but all too often the reverse is true: normally aspirated MR2 owners spend $2,000 for 12–15 whp, possibly compromising reliability in the process.

What to Swap In—Swaps can come in a variety of shapes and sizes. Sometimes the engine comes wrapped in plastic, strapped to a pallet. Other times it is sent as a full clip—the car literally sawed in half from the firewall back (or front, depending on the donor car). Clips are often preferred because the context for the engine is all right there—igniters, hoses, clamps, etc. They are also heavier, an eyesore in the driveway, and more expensive than just an engine by itself. These engines and clips are available because of taxing and registration policies in Japan that make it prohibitively expensive to run the car after a few years (what a convenient policy for the OEMs).

The first step is to decide what engine you want. Kid in a candy store kinda thing. This can paralyze would-be swappers, many unable to decide what exactly they want, and talking themselves into and out of an answer repeatedly before they finally decide. AW15 owners follow logic and often seek out the torquey, ultimately more potent supercharged 4A-GZE from the AW16, though 20-Valve 4A-GE swaps are becoming more common as their prices fall. Others, on a budget and just wanting to get their cars up and running again after an engine failure, might swap in a small-port 4A-GE or another large-port, though as the years drag on, all of these powerplants—small-port and large-port 4A-GEs and 4A-GZEs are becoming less common; there were only so many made.

SW21 owners generally go with Gen II 3S-GTEs. The Toyota/Lexus V-6 option is becoming more common as well, made famous by Brad Bedell and his Ultimate Street Car Challenge bright yellow SW2x. As these swaps occur, how-to information is generously spread over the Internet, making the installs not only more frequent, but often simpler and more refined as well. The information age has certainly changed the face of automotive tuning.

Rebuilding vs. Replacing—Swaps aren't simply for upgrading performance, either. Simply replacing the engine that you've already got can be accomplished through a "swap," though in that case it becomes more of a "replacement." When you lose an engine but want to keep the car, you've basically got two options: rebuild the engine, or replace it.

ADVANCED MR2 PERFORMANCE MODIFICATIONS & THEORY

Engine swaps are a significant undertaking, but can provide years of unstressed, reliable performance. RCrew Racing swapped this Honda K20A2 (200 hp Acura RSX Type S engine) into the ZZW30 MR2 Spyder. Performance should be slightly superior to that offered by a 2ZZGE swap. Photo courtesy RCrew Racing.

Rebuilds, while sounding promising on the surface—an OEM-specification rebuild should leave you with what amounts to a new engine—don't always work out that way.

It is apparently difficult for even the most careful engine builder to reassemble an engine with the same success major manufacturers can. While those same manufacturers consider "hand building" an engine a plus, the same can't always be said for everyone else. Lacking the resources of a major OEM, duplicating their quality, even when sparing no expense in selecting parts, is exceedingly difficult. Rebuilds go wrong with disturbing frequency.

That is not to say that they all go wrong. In fact, done by qualified mechanics, most go right instead. But even a handful of failures out of every 100 rebuilds is way too many, considering the time and expense that goes into a rebuild. The slightest compromise of specification can ruin an engine as soon as it starts: improper head-to-block resurfacing, improper honing of the cylinder walls, improperly sized rod bearings, head stud torqueing not done correctly, extremely slight crank imperfections, and on and on can all ruin your $5000 rebuild.

Dealer-Sourced Short Blocks—If you already own an MR2 Turbo and your stock engine went belly up, an alternative to rebuilding or purchasing a used engine from a clip importer is buying a new OEM 3S-GTE shortblock directly from Toyota. A shortblock is just the bottom-end—the block, cranks, pistons, rods, etc, brand-new and fully assembled to OEM specifications. You'll still need a head, but yours is probably fine, and you can freshen it up separately. Buy a TTE (Team Toyota Europe) head gasket, couple it with a fresh head, and you're in business.

Dirty engines are honest engines, giving you an opportunity to piece together its recent history through spray patterns, locations of grease and grime, oxidation patterns and so on. This clip gives you more information than that sterling picture of the clean engine on eBay.

91

5 TIPS FOR SOURCING A USED MR2 ENGINE

As often as bad rebuilds, I've witnessed swaps go bad as well. After all, these lustful lumps of metal are indeed used, and sometimes have sat for untold amounts of time in Japanese boneyards or American clip importers awaiting purchase. Consider that in lieu of strict Japanese registration and taxing policies, some might've ended up in that boneyard for a reason.

The best engine importers to purchase used engines and clips from are not unlike aftermarket parts retailers, the good and bad ultimately sorting themselves out over the years. Still, when buying a used engine, it is critical that you trust what you're getting. To this end:

1. Consult other owners who've bought engines/clips in the past.

Far and away, this is the easiest way to weed out importers that sell junk. Normally the forces behind the concept of free market will take care of these guys, but sometimes they'll pop up quick and take a whole slew of people to the cleaners before word gets out that they offer junk engines. Also consider the knowledge and reputation of the person (previous customer) offering the feedback; the blind must not lead the blind.

2. Ask to see a picture of the exact engine/clip you are getting.

Some importers won't accommodate you here, but personally I'd like to at least see a photo of what I'm paying $2,500+ for. And a dirty engine is an honest engine. Do not consider a steam-cleaned, sparkling eBay photo as proof of quality—in fact, if it is has been cleaned, it's probably too late to find out anything about the engine from the pictures, so be aware.

3. Ask for a compression and/or leak-down test numbers.

Though impossible to validate long-distance, and not a very reliable indication of the conditions of the internals on older engines, it is better than not having the information at all. Sometimes just asking and listening to the tone of the response can yield some bit of insight that can be helpful in your purchasing decision.

4. Ask for the gauge cluster.

Again, not easy to get, and honestly impossible to absolutely verify its connection to the engine you're buying, but it's better than nothing.

5. Seek out stock engines.

Though the big turbo or exotic exhaust (rear MR2 clips will often have full exhaust systems, being that the entire system is contained within that rear clip) might look tempting sitting there for "free," personally I'd prefer a dead-stock engine to one pounded on in its former life, but that's just me. Certainly some wonderful bonuses can be found in clips, and there's no guarantee on condition or quality in any case.

In the end, an engine swap can be a mixed bag, but if it is big power you're after, or you've got a mint 1986 AW15 and have found the market for clean AW16s lacking, it might be for you. Purchasing your new engine must be done with great caution, and installs must be carefully researched and planned, but ultimately a well-executed engine swap can completely transform your car, offering ridiculous increases in performance and satisfaction of ownership.

Chapter 5
AW11 & ZZW30 Staged Buildups

The AW11 is a great stock performer, and it also still looks downright purposeful more than two decades after it was introduced. Photo courtesy Tommy Guttman.

A PHILOSOPHY OF TUNING

Any modification should be considered an adjustment to achieve a performance level you desire. To do that, you need to identify why the car isn't performing the way you'd like it to, what areas need to be changed to achieve what you want, and the right parts to do the job. Simply throwing parts at a car and hoping it ends up faster is usually a waste money and makes your car slower in the process. Even if you happen to improve its ability to do one thing, you may have compromised another performance sector in the process. Reflect, research and readjust consistently throughout your ownership of your MR2.

When it comes to automotive modification, there is rarely a "best" part or technique. There are too many moving parts, too many possible combinations of parts, too much opinion—thus this staged approach to performance, with specific recommendations as to what you should replace when, and what changes you should expect so you may reorder things yourself as necessary. Ultimately, the most immediate benefit to a staged path to performance is to write out a specific combination of parts and adjustments that work. Individual owners can consider the balance and scope of these changes and can tailor their intent to match what they want for their MR2. This approach should render a well-balanced MR2 that you can then continue to develop as your knowledge and experience increase and as your needs and likes change.

Disclaimer: All of the following parts and procedures are recommendations only. I can't guarantee how they will make your MR2 perform, and some modifications may make your MR2 illegal for the street, depending on the state. Use the information that follows at your own risk.

1985–1989 AW11, MK1
Performance Strengths
- Balance
- Handling stability
- Weight

Areas Needing Improvement
- Horsepower (n/a models)
- Braking
- Seats
- Shifter throw

AW11 Stock Assessment

At around 2400 lbs., the AW15 is relatively light, nimble and a solid all-around performer. Its 0–60 times are generally around 9 seconds, cornering grip is relatively impressive and balance and feedback both well above average. Body roll, in transitions, like all generations of MR2, should be minimal.

Compared to later generations, AW11 brakes are more prone to locking the front wheels during aggressive braking maneuvers. Even though the stock front rotors are vented, the stock brake system is still prone to brake fade when pushed to the limit.

As we've discussed, the stock 4A-GE engine offers modest output, but still begs to be driven aggressively. It has a somewhat narrow range of fun, just north of 4500 rpm to around 7000 rpm, requiring constant attention through the

93

TOYOTA MR2 PERFORMANCE

Being relatively light, mid-engine cars with modest horsepower, most AW15s do not need large aftermarket brakes. Even a modified AW16 can make do with a basic, Stage I brake upgrade under most circumstances. Photo courtesy Hsun Chen.

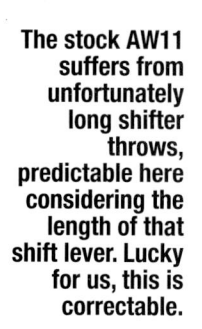

The stock AW11 suffers from unfortunately long shifter throws, predictable here considering the length of that shift lever. Lucky for us, this is correctable.

Being mid-engine, manual transmission MR2s suffer from being cable shifted, with the shifter cables pictured here. These will never allow the shifter-feel of a direct-shifted transmission, like in the Miata.

gearbox to maintain this powerband. This sounds like work, but it renders some of the charm the AW11 is known for. Stock gear change throws are ridiculously long and somewhat sloppy, and the synchronizers don't always age well, especially in the early 1985–1986 models. Overall the gearbox can be considered acceptable.

The stock clutch take-up and overall feel is good, and clutch torque-holding is adequate under most conditions, especially on the later AW11s, courtesy of a larger clutch (212mm, versus 200mm on the earlier versions). The flywheel weight is not particularly lightweight nor heavy, and a good candidate for upgrading in what is basically a sports car heavily dependent on feel.

The non-power steering feel should be tight and direct, though worn bushings may offer up some slop in the steering wheel free-play. Steering wheel shake may be an issue due to bent wheels, or a worn rack and bushings. Ball joints also can wear unexpectedly on the AW11 and must be inspected for wear and specification.

The seats, were probably the most significant of the interior upgrades in the second-generation MR2, and the increase in function and aesthetics they offer should not be ignored. The stock seats in the early AW11s are functionally decent, but not great for serious cornering. Side-bolstering and available adjustment is average at best, and visually they range from truly ugly (red and blue/black early seats) to bland (1988–1989 seats seemingly torn right out of a 4-door Corolla).

The steering wheel is not particularly attractive, nor does it host an airbag, so unless you are trying to maintain a dead-stock AW11, it's a good place to upgrade.

Toyota also revised the gear change and rear suspension during AW11 revisions, and obviously offered more power with the supercharged MR2 in 1988. The AW11's most significant weakness doesn't involve dynamics of performance—it's corrosion.

AW11 STAGED BUILDUPS

A staged tuning guide is a logical and efficient approach to building a performance vehicle. This method also helps illustrate what modifications are necessary at what point, which modifications require others, and how to prioritize and plan accordingly.

Such a plan also needs to be reflected on with a basic understanding of the reasoning behind each modification. If you can grasp the fundamental motivation behind the modification, it will be easier for you to make your own plan accordingly,

AW11 & ZZW30 Staged Buildups

Tommy Guttman's turbocharged AW11 looks fast just sitting on the page, doesn't it? Though the T-top roof is less than ideal, somewhat compromising chassis rigidity, Steve has developed the car so well through extensive track-testing that for all intents and purposes, the roof is a nonissue. Photo courtesy Tommy Guttman.

Counter-steer is steering into the direction the back of the car is heading in an attempt to correct a non-ideal on-course path. Drifting, demonstrated by this AW11, is an exaggerated practice in this skill. The MR2 is less frequently drifted than other rear-drive sports cars because its relatively short wheelbase and mid-engine design make it more challenging to drift. Photo courtesy Mays/Heitkotter.

rearranging the buildup of your own MR2. It will also assist you should you decide to upgrade again in the future in certain sections—suspension only, for example—so that you do not see a particular stage as an all-or-nothing proposition. In this way, though it is usually recommended, Stage III does not necessarily require every single Stage I modification.

Stage 1
- Tokico Illumina shocks
- Alignment
- Short-shifter and shifter bushing
- Porterfield R4S brake pads, ATE Type 200 fluid
- Redline MT-90

Adjustable Shocks—Added dampening force and adjustability, even on stock springs if your MR2 is still so equipped, is our first order of business. Adjustable shocks can not only help reduce wasteful suspension movement, but can also help get the AW11 to turn. The stock AW11 tends towards understeer. There isn't enough power to adjust cornering attitude with the throttle, and the AW11 very much likes to be driven by the scruff of the neck. Adjustable shocks on stock springs should be the bare minimum for any performance car, and should be the first step of a Stage I buildup.

While Koni makes exceptional shocks, you need to think long-term. If you're never going to go further in your shock-and-spring setup than a set of adjustable shocks, Konis are worth the money. If you are leaning towards a coil-over purchase later on, you might want a less expensive set of shocks, or hold off on shocks all together. For all-around bang-for-the-buck, the adjustable Tokico Illuminas shocks are hard to beat.

The AW11 interior, revealed here by this clean AW16 with optional leather interior, is functional, dated and minimalist. Still, there is room for improvement, depending on what your objectives are—aesthetics, authenticity or performance. Photo courtesy Helder Carreiro.

Getting your MR2 OEM healthy applies to suspension, too. This is especially critical with the AW11 and SW2x, which tend to be hard on ball joints, and sensitive to out-of-whack alignments, so check these items.

Note: The MR2 actually uses a MacPherson strut design, with strut inserts, but the aftermarket term is shocks, so in the interest of common language, that's what we'll call them.

Toyota MR2 Performance

RECOMMENDED ALIGNMENT SETTINGS		
	Street	Track
Camber Front	-1 degree	-2 degrees
Camber Rear	-1"	-1.5" degrees
Caster	+6	+6
Toe-in Front	1/16"	1/16" (1/8" out for autocross)
Toe-in Rear	1/8"	1/8" (1/16" out for autocross)

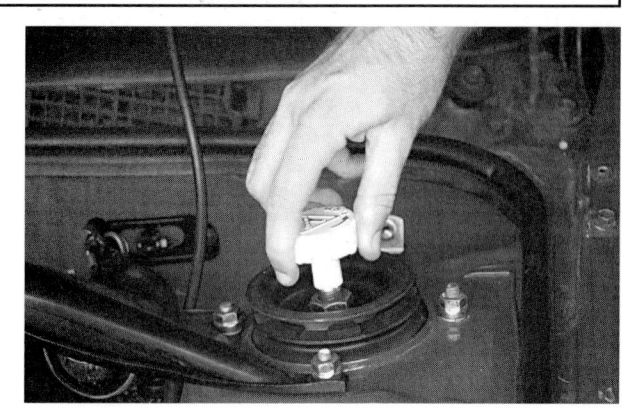

Adjustable shocks, like the Koni units on this 1987 AW11, should be considered essential equipment on almost any sports car. You can also just make out a stock front strut tower brace at the very left. Photo courtesy Bryan Heitkotter.

If you look closely, you can see the negative camber in the rear of this AW11. It is set at -1.5". Photo courtesy Bryan Heitkotter.

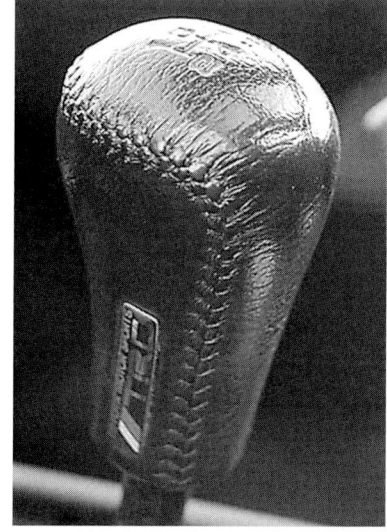

Choosing a shift knob should be done by shape and feel. Early AW11s shipped from the factory with a banana shift knob, which is easily replaced. This one from TRD actually improves both aesthetics and shifter feel, requiring no compromise either way. Photo courtesy Bryan Heitkotter.

Alignment—Almost two decades ago now, TRD (Toyota Racing Development) released the following suggested alignment specs for the AW11. Generally speaking, the biggest difference between the two settings is increased negative camber in the front, which should help increase front grip, at the cost of some front tire wear (you can help minimize this by rotating your tires often).

How your car handles depends on a variety of factors, and this is very important to consider. Tire pressure, tire compound, roll stiffness, driving surface, roof-type, driving style, and countless other factors produce the "handling" your MR2 is capable of.

I recently tested a 1985 MR2, along with Tokico shocks, stock springs and R-compound tires, ST sway bars, and Prothane Bushings, and had excellent results with the above "Track" alignment, but with a bit more camber up front (2.5"), and in the rear (2.0"), and 0 (zero) toe in the rear. The result was very neutral handling (as long as the tires had enough heat in them), with very slight initial understeer, and oversteer available. High-speed stability was not compromised, and braking was stable. The car was able to put down admittedly meager power well out of corners, and had accurate and direct turn-in.

Remember, experiment with alignment settings until your MR2 reacts how you want it to, in a way that best suits your driving style. Alignment is not a set-it-and-forget-it kind of thing, and requires attention as components wear, bumps are hit, and your use of the car changes. In fact, take advantage of every adjustable setting your suspension has until you achieve the handling balance and feel you desire—alignment, sway bar settings, shocks settings, and tire pressure. The above simply constitutes an appropriate range to begin fine tuning your alignment.

Note: Try to find a local shop that specializes in "autocross" or other similar high-performance alignments. Though your local tire store may indeed have an alignment rack and someone they pay to use it, don't assume your "lifetime alignment" agreement means a lifetime of optimal camber, caster, and toe. A good alignment shop is worth its weight in gold, and expensive rubber.

AW11 & ZZW30 Staged Buildups

Replacing the stock rubber shifter bushing with one made from brass, which won't deteriorate or deform under movement, can improve shifter feel tremendously. How much depends on the condition of your stock bushings.

These pads are one-third of what you need for more consistent high-performance braking from your MR2. If you look closely, you can see where the braking piston impacts the pad, forcing it to squeeze down on the brake rotor. These pads were used for track events, then taken off for daily driving. (Incidentally, the other two-thirds are brake fluid and tire compound.)

Short-Shifter, Shift-Knob and Shifter Bushings—Even the later, revised C52 transmission can only be shifted as fast as the synchronizers allow, regardless of shifter length or height. The motivation here is not necessarily faster shifts, but more direct shifts featuring better feel.

Stock AW11 shift throws are long and after many years of use, likely loose and sloppy too. A good place to start is Twosrus, an MR2-specific online vendor that carries aftermarket shifter bushings and a shifter lowering plate. Being a cable-shifted car, the cables connect to the transmission by lever via a rubber bushing, and, as we know with suspension bushings, rubber deteriorates over time. Twosrus offers a solid brass replacement that improves shifter feel and accuracy greatly. In my 1985 MR2 with 93,000 miles, after the brass bushing installation, the shifter felt as though it were connected to something solid and whole when it was moved—something specific—rather than simply stirring oatmeal. The combination of Jon O. shift kit, RSe lowering plate and SpeedSource bushing kit will vastly improve the shifting feel and throw length of your AW11, for about $60, and the install is a breeze with provided instructions. Not bad. Finishing everything off with quality shift knob will give you a gear-change setup that will be vastly improved over stock, for very little investment.

Do note, however, that for whatever reason some folks have issues engaging gears after installing these bushings, with one owner I know (who eventually had to go back to his factory bushings) theorizing that the flexibility of the factory rubber units might actually be useful under certain conditions when the suspension is highly loaded, and the engine is twisting in the engine bay under its own torque, with said movement changing the alignment of the shifter cables and the transmission itself.

Brake Pads & Fluid—Everyone chases big brakes—they do look cool after all—but big brakes generally are not necessary, especially considering our relatively lightweight and meager horsepower levels at this point. The question we must ask here is, as always, what is the car not doing that we want it to do? What exactly are we trying to improve? Stopping distance? Brake fade? Brake modulation? In reality, all brakes do is convert the kinetic energy of the car in motion to heat through the process of friction.

Sticky tires, quality performance brake pads, and fresh, well-bled performance brake fluid will go a long way in providing your AW11 with capable, consistent stopping. Do not underestimate the potential of this setup.

As mentioned, the front wheels tend to lock problematically earlier than the rear. To help prevent this, it is helpful to use different brake pads front and rear, choosing front pads with less friction compared to the rear. Gaining this kind of information can be difficult—which pad is has sufficiently reduced friction compared to the front?

Fortunately, the relatively low cost of brake pad allows for some experimentation, and aftermarket pad manufacturers (or vendors) are often more than happy to discuss brake pad specifics with gearheads like ourselves trying to create a hybrid setup. This is also where message boards come in handy.

Portfield R4S, Carbotech SSF and Panthers and Hawk HP+ are all excellent brake pads with AW11

TOYOTA MR2 PERFORMANCE

Brakes are a system. High-performance braking is mix of tire grip, fluid boiling point and friction between the brake pad and rotor. Performance compound tires like these Falken 452s will immediately increase braking performance all by themselves. Photo courtesy Falken Tires.

applications available, though the Hawks especially are very noisy. Remember that although increased clamping power can be helpful, the main goal is to reduce brake fade. Coupled with sticky tires, braking distances should be decreased anywhere from 10' to 30' or more, depending on how bad they were to begin with.

Note: Though Toyota began adding larger brake rotors and calipers in 1987, the above modifications are no less useful to those models. In lieu of considerable size upgrades, brake pad composition and brake fluid properties are the most critical players in avoiding brake fade.

New OEM Tie-Rods and Balljoints—Though this is redundant since we consistently stress any modified car should be at least OEM healthy, these are high-wear items on MR2s, and their lack of proper function can be misdiagnosed as something else, a worn steering rack or bent wheels, for example. Be sure yours are in good shape, and if there is any question, replace them. Twosrus is an excellent resource.

Stage II
- Sway bars, sway bar end-links
- Strut tower braces
- Aftermarket seats
- Lightweight flywheel
- Stainless steel brake lines, master cylinder brace
- Lightweight, relocated battery

Sway Bars, Sway Bar End-Links—A set of ST sway bars and end-links from Twosrus is the way to go here. As an alternative, you can simply hunt down an OEM rear sway from an AW11. (Though SW2x sway bars fit, their stiffness and front-to-rear ratio does not translate well to the AW11.)

It may be necessary for you to purchase new sway bar tabs before installing your sway bar—especially if you've simply located a used 1985 rear sway bar instead of purchasing a full set—and finding that out before you place your order is a good idea. Stick your head underneath your MR2 and look where

Toyota stopped installing rear sway bars sometime in early 1986. Most MR2s manufactured between 1986 and 1989 do not have the sway bar tabs in the rear necessary to install a sway bar, but lucky for you, Twosrus keeps these parts in stock. Oddly enough, some MR2s did receive the tabs but no sway bar. Photo courtesy Mike Choi.

Side by side, the beefy aftermarket end-links from Twosrus are immediately distinguishable from the one on the left top. Photo courtesy Mike Choi.

the sway bar should mount between the rear suspension arms. If there aren't any tabs there—look hard, they might just be dirty—you'll need them.

When Toyota delivered the AW11 to the U.S. market it did so with a rear sway bar. For some reason though, Toyota stopped including a rear sway bar sometime in early 1986. It was pencil-thin, but theoretically a perfectly balanced car should require no sway bar at all, so let's consider that a compliment. The sway bar provided the MR2 with a commendable handling balance, but was dropped sometime soon thereafter, and didn't reappear again until the AW16 was born at the end of the AW11's run. It is not entirely clear why

AW11 & ZZW30 Staged Buildups

This AW16 has a factory front strut tower bar, so if an aftermarket strut tower bar proves impossible to course, or is beyond your budget, there are alternatives. Photo courtesy Helder Carreiro.

The 1987–1989 stock seats aren't too bad, but lack the kind of side-bolstering a modified AW11 requires. 1987 AW11 seats are pictured. Photo courtesy Bryan Heitkotter.

The side bolsters on the stock leather seats seem to hold their shape well over the years, but the leather surface allows a bit more movement during aggressive maneuvering than is ideal. Photo courtesy Helder Carreiro.

ZZW30 seats in an AW11 is a fairly straight-forward swap, provided you know a professional who can do some very light welding for you. The brackets and mounting anchors must be exchanged. Photo courtesy Bill Merton.

Toyota dropped the bar, though it more than likely has to do with a confluence of oversteer, bad drivers, and litigation.

Note: Those on a budget can try to track down a stock sway bar from a 1985 AW15. It doesn't make a dramatic change in roll, but will improve the balance—the ratio of roll stiffness between the front and rear of the car. Where do you find one? They can surface from time to time on MR2 message forums and eBay.

Strut Tower Braces—Cusco and GReddy have manufactured some pieces, though these can be difficult to locate. Google and MR2 online forums can help here. Overall, these can be difficult pieces to source for the AW11.

Later AW11s did come with front strut tower bars, so there is a cheap source of a decent bar there if you can find it.

Improved Seats—As with many other modifications, the time to perform this one is subjective. An AW11 with a set of grippy tires, adjustable shocks and a good set of aftermarket seats offers surprisingly satisfying performance, and has the benefit of being modest and affordable. Aftermarket seats are often postponed by enthusiasts looking to avoid superficial modifications that only give the appearance of speed.

The real measure of a modification's necessity is best made against an evaluation of the stock piece's performance—and as we mentioned above, the AW11 seats aren't great. Installing replacement OEM seats—for example, a set of clean, early 1985

99

A quality aftermarket seat is among the nicest additions you can make to any MR2, especially an AW11. Rather than absolute, dominating performance, the AW11 is about a pure driving spirit, and a seat from Bride (pictured), Recaro, Sparco or Momo can enhance that feel tremendously. If you're on a budget, replace one at a time. Photo courtesy Brian Hill.

The brake master cylinder can move slightly under aggressive braking. Some method of bracing the master cylinder against such movement will return improved pedal feel. Photo courtesy Jensen Lum.

Stainless steel brake lines will enhance pedal feel by the reduced compliance of the stainless steel braided line. How much it improves pedal feel and your ability to modulate braking depends on the condition of your stock lines.

AW11 seats—can be as simple as removing each of the four bolts that anchor the seat to the floor, disconnecting the seat belt sensors or other relevant safety wiring, and swapping in the new seat.

But most aftermarket seats require the use of aftermarket brackets and sliders from manufacturers such as Wedge. These new brackets can also sit you higher in the cabin as well, which can ruin some of the excellent feel of an MR2 cabin.

As for ease of ingress/egress, I am generally unconcerned—compromises must be made to some degree—but for many owners this is an issue. And while the added function, aesthetics and lighter weight of a good set of seats is unquestionable, skip the eBay, no-brand specials unless you can sit in them and are sure they'll not only fit, but last. Don't skimp on seat quality, especially since they are so critical to safety.

Note: Before purchasing an aftermarket seat, be sure it has the kind of adjustments you require. Typically those most affordable aftermarket seats have little to no adjustment at all, and though this is what ultimately contributes to their extremely light weight, be sure this is something you can live with.

Lightweight Flywheel—A lightweight flywheel will not yield dramatic gains in performance, but as we've discussed, on an AW15, short of major mechanical efforts, little will. It will, however, increase the spirit and feel of the car, which is what an AW15 is all about anyway, allowing quicker revving and easier rev-matching while you carve through your local canyon or country backroad. A lightweight flywheel should slightly reduce 0–60 times by a couple of tenths of a second, and making the car's acceleration feel more immediate. The improvement in engine responsiveness here is worth the effort. Fidanza makes an excellent aluminum flywheel for the AW15.

Stainless Steel Brake Lines and Brake Master Cylinder Brace—In addition to our pad/fluid upgrade in Stage I, stainless steel lines such as the quality units available from Goodridge should firm up our brake pedal some and hopefully allow for less frequent locking of the fronts by allowing you to better gauge appropriate levels of pedal pressure under different braking circumstances.

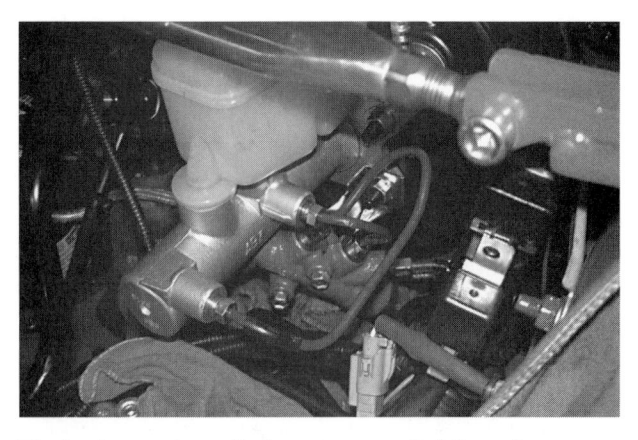

Stainless steel brake lines are not as urgent as a proper alignment for example, but considering their relatively low cost, the age and probable condition of the stock brake lines on any AW11, they're cheap insurance.

The brake master cylinder visibly moves when aggressive brake pressure is applied. This results in a loss of true input and feel, but can be avoided with the use of a brace. A company called High & Tight used to manufacture such a brace, but it has been discontinued. As with many other AW11 modifications, your only option is to go custom. Get creative, and figure out a solution that disallows master cylinder movement. Necessity is the mother of invention, after all. At this stage, reduced-fade braking and consistent pedal feel are our goals.

Lightweight Battery, Battery Relocation—In its evolution of the MR2, Toyota moved the battery from the engine bay in the AW11 to the front trunk in the SW2x. With an already pronounced 45/55 rearward weight bias, moving 35 lbs. (stock battery weight) of battery to the front not only cleans up the engine bay some, but takes some weight off the already heavy rear. It also allows for more liberal use of engine bay space when considering other modifications like cold-air intakes or turbochargers. Also consider a lightweight battery, such as that available from Optima, or even a stock Mazda Miata—just be sure to properly secure the battery.

Note: Tempted as though you may be, it is not advised that you run a power wire through your MR2's cabin when performing this modification. A safer alternative is to run it underneath the car, suitably isolated from potential snags and debris by some method of cover. Also obtain a proper battery enclosure for your relocation, and use substantive (low gauge) wire, and fuses. Most relocation kits have everything you need to do it right.

Stage III
- Wider, lighter wheels
- Urethane bushings
- Custom K&N air filter setup, OEM AW16 muffler
- Special: 1987–1989 OEM brakes
- Engine swap

Wider, Lighter Wheels—The AW15 factory wheel is 14x5.5, and though they can accommodate tires as wide as 225, this makes for a very tall, mushroomed sidewall, and is far from ideal. The wider a tire is, the bigger the potential footprint or contact patch with the road, but a 225 tire will stand taller on a 5.5" wheel than it would on a 7"—so wider doesn't always necessarily translate to an equally wider contact patch.

Toyota has increased the wheel sizes of both the SW2x and ZZW30 over the course of each generation, but oddly never did so with the AW15, only adding slightly wider wheels when adding the supercharged engine.

The AW15 came equipped stock with 185-series tires, but wider tires can be used as well. Ideally, a

Lightweight batteries like those available from Optima can be an inexpensive, no-compromise way to shed a few pounds. Optima batteries also feature lower rates of self-discharging, and generally extended life over conventional batteries. Photo courtesy Optima Batteries.

Wider, lighter wheels, like these from Rays Engineering/Volk, allow you to place a wider footprint on the road and lose rotational mass as well. They marginally increase acceleration, braking and transitional response during handling maneuvers. Photo courtesy David Jones.

195/50/14 size tire would work well on the stock wheel, but tires available in that size are few and far between. Due to the questionable aesthetics of every stock wheel Toyota ever bolted on to the AW15, simply replacing the wheels altogether with a set of lightweight (remember, rotational mass), 15x7 wheels would allow not only a wider footprint, but a wider selection of high-performance tires.

Though you might be tempted to go with an excessively large wheel, your best bet for most applications is a 15" wheel. Tire availability won't be an issue, several manufacturers like Kosei and

Early Mazda Miata wheels on an AW11: the poor man's Panasport, but sadly no wider than what you got on your MR2 stock.

Reduced compliance is the big idea with aftermarket suspension bushings, leading to sharper handling, and the potential for a more optimal alignment with the road as well. Photo courtesy Mike Choi.

ZZW30 wheels offer an even better option than Miata wheels, being larger and staggered front to rear for increased rear handling stability, exceptionally lightweight and of Toyota MR2 DNA. These are also a good alternative for those modifying on a budget.

Rota make inexpensive, lightweight 15" wheels in MR2 offsets, and the sidewall won't have to be spaghetti-thin, allowing for a wheel-and-tire package that is affordable, lightweight and easy to drive at the limit.

Wheel styling is so subjective that recommending one set over another for any reason other than weight, width, durability or cost doesn't make a lot of sense. That said, the Kosei K1 is a fairly lightweight, inexpensive wheel if you want some advice on where to start. Enkei, who manufactures several wheels for OEMs—Honda and Mitsubishi for example—also manufactures good, affordable wheels.

Polyurethane or Hard Rubber Suspension Bushings—Worn factory bushings allow for undesirable changes in suspension alignment, and yield a sloppy, indirect feel through the steering wheel. New factory bushings would be a big improvement over worn units, and hard rubber bushings, as available from the TRD suspension bushing kit, offer even less compliance. Even less compliant are Prothane bushings. Twosrus or any other mainstream vendor should carry Prothane bushings.

Years ago, TRD made a set of suspension bushings out of hard rubber, the idea being that hard rubber is more firm than OEM bushings while not suffering from some of the criticisms of urethane bushings, like sometimes binding, or being noisy. These bushing sets are no longer readily available from traditional vendors, but if you stumble across a complete set, they are a worthwhile upgrade.

Intake and Exhaust—Generally, intake and exhaust modifications are among the first approached by the tuning owner, but as we covered in Chapter 3, on the AW15 their role is far less concrete than you might think. In fact, I hesitate to recommend them at all unless you've installed aftermarket camshafts and cam gears, or added some form of engine management.

In a nutshell, getting air into and out of the engine is critical, but since the factory airbox isn't a significant "bottleneck" at stock or near-stock

AW11 & ZZW30 STAGED BUILDUPS

The bad part about the suspension bushing upgrade is the install. It is not a fun Saturday off from work installing these bad boys. Hammers, chisels, presses and even blowtorches may be necessary. Photo courtesy Mike Choi.

The supercharged exhaust will not simply bolt on. The B-pipe must be modified, exactly how depending on the size exhaust you are running. A muffler shop should be able to perform this quick job on the cheap.

The OEM exhaust from the AW16 has a more aggressive look than the normally aspirated version, is inexpensive and sounds marginally deeper without being the least bit offensive. It also should be good for a couple of extra horsepower.

power levels, adding an aftermarket air filter isn't going to add significant power—there simply isn't a significant need for extra airflow.

For the AW15, a stock muffler from the AW16 is available for around $100 from your local dealer. The muffler tips don't come with the muffler, and are almost $50 each from the dealer, quickly doubling the price. If you can track down a set of used tips, you'd be better off. This is a solid choice for AW15 owners.

As for the air filter, if you must upgrade here, you're going to have to go with a custom setup, though with 4A-GE (AW15) intakes this can be as simple as replacing the airbox and intake tract from the airbox to the throttle body.

A "Stage III" AW11 is an adjustable, stable, well-balanced MR2. Balance more than dominance in any single area is what MR2s were engineered for, and represents our approach to our modifications accordingly. Photo courtesy Hsun Chen.

103

Toyota MR2 Performance

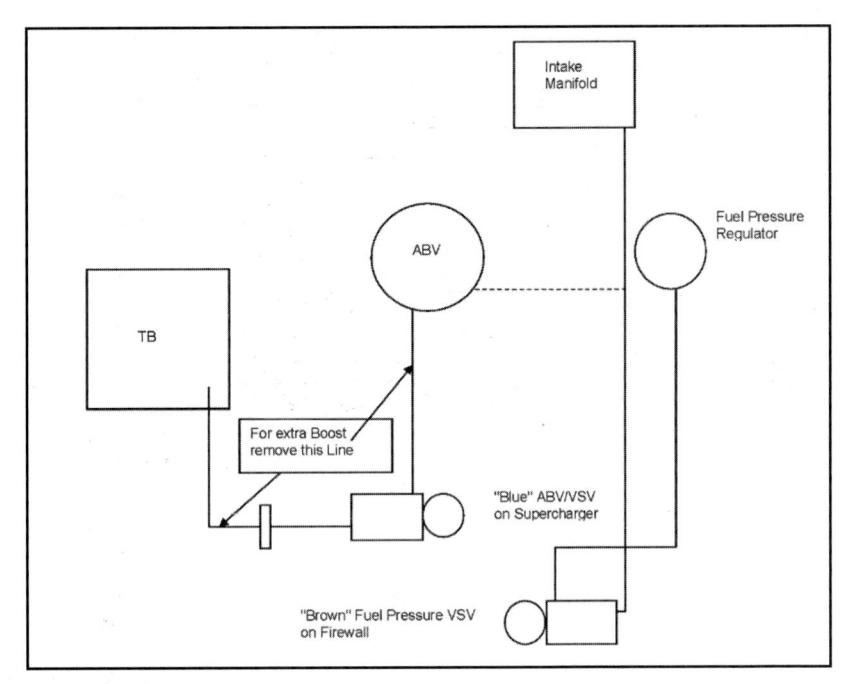

To increase the boost on a 4A-GZE, you must first deal with the ABV. Remove the line on the throttle body which comes from the blue ABV/VSV. Plug the fitting on the throttle body. Next, pull the line coming from the ABV to the blue ABS/VSV, and "T" it into the line between the intake manifold and the brown fuel pressure VSV (shown as a dashed line). This will allow intake manifold pressure to help keep the ABV closed.

1988–1989 SUPERCHARGED MR2

Overall, the same path of development applies to both the AW15 and AW16, with minor changes for the overall car as a platform, but obvious revisions considering their unique powerplants. There are also other minor issues, such as the reality that the AW16 has a intercooler that gets in the way in the engine bay, necessitating a unique strut tower bar (Cusco has manufactured such a piece). Also, the forced-induction should render slightly different priorities, namely a newfound focus on horsepower, as well as an attention to durable, repeatable performance, and ways to help exact that, like water injection, the Grunt Box, and so on. Exhaust, too, is more of a priority on a forced-induction engine.

Right away, the AW16 is faster. Stock versus stock, the 1988–1989 supercharged MR2 is about 2 seconds faster to 60 mph, and about 1.5–2.0 seconds faster through the quarter mile than the AW15, and that can be interpreted two ways: that it's quick enough that we needn't increase horsepower just yet, or the fact that its faster, and forced-induction horsepower comes so much easier, that we should start there. In our staged process, as usual, we are going to enhance the durability of our stock setup before we start leaning on it.

Consider the following engine modifications for

The AW16 is basically an AW15 with a supercharger to boost mid-range torque, so the overlap in suggested modifications between the cars is significant. Photo courtesy Helder Carreiro.

the AW16, this being in addition to the standard AW15 changes noted previously.

Exhaust Gas Temperature and Boost Gauges—GReddy, HKS, Blitz, Omori, Autometer, etc., all make quality gauges.

Water Injection—We've covered the benefits of water injection elsewhere, and since the AW16 is forced-induction, it can join in on the fun, unlike the AW15.

ABV Modification—The "Air Bypass Valve" is sort of like a blow-off valve on the MR2 Turbo, only not. Instead of venting boost once the throttle is closed, it is used instead to help bleed off boost once it sees the factory-set 8 psi, and to route air around the supercharger when it's not spinning.

To run more than stock boost, which is what we're after, we need to first take the ABV out of the picture. To do this, we must route the vacuum hose from the ABV to the intake manifold, and cap the line that ran from the VSV to the ABV. Another option is to remove the ABV altogether, though the long-term effects and success of this method are unclear, and it is not recommended.

Grunt Box—4A-GZEs tend to run lean under 3500 rpm, a problem only exacerbated when the boost is increased. To combat this, an ingenious, simply designed piece of hardware called "the Grunt Box" can be purchased for around $125.

The Grunt Box utilizes the cold-start injector to supplement the fuel delivered in high-load, low-rev situations—that is high-throttle input, and low rpm, reducing or eliminating the lean condition.

The results of this added fuel are not just safety-related, either. Mike Gruber and Rob Files, the Grunt Box's manufacturers, claim a 10 lbs-ft. increase on a Cusco pulley-equipped supercharged MR2, and improved throttle response, though the exact improvement on your MR2 depends on your setup. The idea here, though, is safety margin and durability.

AW11 & ZZW30 STAGED BUILDUPS

The 4A-GZE, pumping considerably more air than the normally aspirated 4A-GE, will make proper use of a header. This 4A-GZE is hiding a Cusco header beneath some very expensive heat shielding that is used to help the exhaust manifold to retain its heat, promoting exhaust scavenging.

Header/Exhaust—Any time you turn up the boost, factory exhaust piping quickly becomes restrictive. Your best bet for an exhaust is to go with a custom setup (like the AW15), though in the case of the 4A-GZE, go for a 2.5" exhaust. Unlike the normally aspirated 4A-GE, a header on the forced-induction 4A-GZE is more worthy of your efforts, not only in reducing weight, but also aiding in the age-old endeavor of getting used air out of the engine. Remember, as horsepower levels increase, the potency of crucial airflow modifications increase as well.

More Boost—The way to turn up the boost on a supercharged car is by installing an oversized crank pulley. The factory supercharger pulley is 145mm, and the Fensport kit, for reference, is 176mm, the NST kit 180mm. These larger pulleys provide a modified drive ratio to force the supercharger to produce additional boost. If your 4A-GZE and clutch are healthy, this is a wonderful thing, upping the torque considerably.

These modifications to the pulley size differ by manufacturer, as do the corresponding boost levels provided by the kits. A problem here is going to be actually getting your hands on one of these kits. Many different 4A-GZE supercharger pulley kits have come in and out of production, and being such a small (and shrinking) market, most major aftermarket manufactures don't bother with development of new products for the MR2, especially the even less common 4A-GZE-powered AW16. For products like a supercharger pulley,

4A-GZE SUPERCHARGER PULLEY KITS BOOST LEVELS

HKS:	10 psi
Blitz:	12 psi
Fensport:	12-14 psi
NST:	13.5 psi
Cusco:	14 psi

The Fensport 4A-GZE pulley kit comes with an undersized idler pulley and new drive belt as well. The Cusco kit also has a built-in harmonic dampener like the factory unit. Along with a header/exhaust combination, such a kit should put you in the neighborhood of 14 psi, up from the factory 8 psi, together offering a 25–35 whp increase.

Note: More aggressive tuning, i.e, boost levels, timing, etc., often require spark plugs "colder" than stock. Fensport recommends exactly that for their 4A-GZE pulley kit.

where a custom piece isn't feasible, community-based marketplaces like Toyota and MR2 message forums, and eBay as well, can often turn up product.

Note: Other Toyota models, namely AE86 and AE92 Toyota Corolla variants, often swap in 4A-GZE engines, and in this way can be an excellent source of technical and product information. www.club4ag.com is an excellent resource here.

Beyond

There is no bolt-in option for an aftermarket intercooler, so a custom, high-efficiency intercooler would be worth the effort if you're looking to go beyond the standard exhaust-and-pulley AW16 combination.

The Megasquirt is an oddly named but effective do-it-yourself programmable EFI controller, composed very simply of a processor, main board and associated code and software. No engine-

While somewhat compromised because of some extra weight and a less-rigid T-top roof, the AW16 nonetheless has something the AW15 does not—instant, line-correcting torque. A Stage III (and then some) AW16, nicely represented here, should offer vice-free handling, and excellent mid-range torque to make corner-exits consistently quick. Photo courtesy Tommy Guttman.

Toyota MR2 Performance

A brake proportioning valve, such as the one pictured from Wilwood, allows for brake bias to be adjusted from front to rear. Since the AW11 likes to prematurely lock the front brakes, this modification can help reduce or altogether eliminate this tendency.

management system is simple, but the Megasquirt not only provides an affordable method of engine management, but your own level of efficacy in dealing with that engine management as well. It is very much a learning tool as much as a horsepower tool. Personally, I'd rather not "learn" on something that is going to cost two or three grand to replace should that learning be slow in arriving, but many users have found the Megasquirt's affordable simplicity the perfect match for their equally simple, and relatively inexpensive modified AW16. As the complexity of your setup increases, so does the necessity for "smart" engine management capable of dealing with your modified 4A-GZE.

Beyond the adjustment of ratios of pad friction from front to rear mentioned in Stage I, a brake proportioning valve can also be plumbed into the stock system to help control the lock-up of the front wheels, though the potential consequences of a failure of such a modification relegate it to the more advanced MR2 development. Removing any brake-dust shields and adding brake-cooling ducts would also further enhance the fade-resistance of the brakes, and for little to no money. Also, the stock clutch, on both the AW15 and AW16, should be fine at stock or near-stock power levels. Once you start radically increasing torque—most likely on the 4A-GZE-powered AW16—you will need to consider an aftermarket clutch with increased torque-holding capacity. As general advice, though, I would not replace the OEM clutch until you are experiencing a healthy OEM clutch not able to handle what your engine is producing. The quality of aftermarket parts in general, especially clutches, is often suspect.

As with all MR2s, the car can be lightened, camshafts can be replaced, headwork can be performed and even the supercharger itself can be replaced by larger units. These modifications, though, cost a lot of money when you add them all up. When you consider modifications as adjustment of a system, rather than independent objects to lust after, it helps keep your budget and overall planning modest, and keeps your focus on maximizing your opportunities to drive your MR2 competitively, for recreation and everything in between.

2000-2005 SPYDER
ZZW30 Performance Strengths
- Weight
- Balance
- Braking

ZZW30 Areas Needing Improvement
- Chassis stiffness
- Horsepower
- Engine reliability (pre-cat failure)

ZZW30 Stock Assessment

The stock ZZW30 owes much of its athleticism to its light weight. At 2200 lbs., it is around 200 lbs. lighter even than a sparsely-optioned AW15, while maintaining a power to weight ratio closer to the hulking SW20 than the AW15, and acceleration times somewhere between the two.

This low weight allows the ZZW30 to literally do everything easier: brake, handle and accelerate. Its braking from 60 mph was generally reported in under 120 feet, and lateral grip close to 0.90g's. Even with the relatively meager horsepower offered up by the 1ZZ-FE, because of that weight the

The MR2 Spyder, out of the box, is the best-handling MR2 ever built, thanks in no small part to its 2200-lb. curb weight. Photo courtesy Toyota Motor Sales.

AW11 & ZZW30 Staged Buildups

Its lightweight, mid-engine setup and rear-drive design are its best features. A modest engine, weak structural rigidity and lack of rollover protection are prime areas for improvement. Photo courtesy Toyota Motor Sales.

The transitional response of any MR2 should be just short of brilliant, with the low polar moment of its mid-engine design, and relatively lightweight, only somewhat let down by its simplistic MacPherson strut design. Nonetheless, the ZZW30's stock handling performance is stunning and has even greater potential. Photo Hsun Chen.

ZZW30 is quick, its 0–60 mph time around 7.5 seconds.

The steering is direct and quick, but the steering feel is artificially numb compared to the non-power-assisted AW11. Being considerably younger than the AW11 and SW2xs, the ZZW30 doesn't suffer the same effects of old age that can cling to its predecessors: there should be less urgency in replacing shocks, suspension bushings, and other rubber/plastic/wearing parts. The brakes, owing to ABS and EBD, should simply feel better right off the bat, and items like seats, steering wheels, and other interior appointments should all (generally) be in better condition as well.

1ZZ-FE Pre-Cat Failure

In the name of reduced emissions—and to get the government off their back—Toyota installed pre-catalytic converters on their little roadster to supplement the standard emission-reducing equipment. The key difference between a pre-cat and a, well, non-pre-cat, is location: pre-cats are placed right off the exhaust manifold. The idea behind these pre-cats is that by being placed as close as possible to the exhaust ports in the head, where things are much hotter, they can reach operating temperature faster. Engines like to be warm. Just as significant portion of the wear in the combustion chamber occurs when the engine is "cold started," emissions, too, suffer when things are all cold.

Very basically, a catalytic converter is a device placed in the exhaust stream that is able to reach and sustain—hopefully with great durability—extremely high temperatures. At these very high temperatures they are able to convert the hydrocarbons, carbon monoxides, and nitrogen oxides—natural by-products of the combustion process—into less harmful versions of their former selves. The problem arises when these converters, located so close to the cylinder head, where a staggering amount of air is processed, begin to break into little pieces. Ever laugh while eating a cracker?

You might assume that the exhaust would blow the debris out, not suck it in, and it usually does, but through a phenomenon called reversion, the exhaust can sometimes be drawn back into the engine, and if there are loose bits of pre-cat within breathing range, they could be ingested back into the engine.

Via exhaust reversion, which happens when exhaust gases are

Adjustable suspension pieces are crucial to really optimizing the performance of your MR2. Koni, Tokico and KYB (pictured) all offer quality aftermarket shocks/struts. Photo courtesy KYB.

107

Toyota MR2 Performance

To reduce emissions—and to get the government off their back—Toyota installed pre-catalytic converters on their little roadster to supplement the standard emission-reducing equipment.

Placing them close to the exhaust valves allows them to light—and begin the emissions-altering process—more quickly.

The problem occurs when the material the converter is made of begins to break apart. If you are able to pull the manifold/converter completely off the car, you'll have a better vantage point to locate material debris.

temporarily sucked back into the head, little bits of a pre-cat may enter the combustion chambers (assuming the pre-cat is failing). When foreign objects enter the combustion chamber, things go bad in a hurry. How often do they fail? Not often enough for Toyota to issue a recall on them. However, in a recent poll within an MR2 community, scores of Spyder owners reported failed pre-cats and subsequently grenaded engines, with several of those reporting multiple failed engines. With the MR2 Spyder no longer in production in the United States, the chances of a recall are probably diminishing rapidly, and to be fair, mathematically the percentage of failures still isn't very high. But try telling that to the owner of an MR2 Spyder sitting on a lift without an engine.

Three Checks for Pre-Cat Failure

Check #1: Visually Inspect the Pre-Cat Material
1. Remove the O_2 sensors.
2. Visually inspect the pre-cat material—look for any bits of loose debris. The catalytic converter failure is physical—it will physically break up, but won't always be simple to spot. Be sure what you are actually able to see is completely intact.
3. Reinstall the O_2 sensors.
4. Start the car to make sure you don't have a check-engine light.

Check #2: Look for Exhaust Debris
1. Place a white plastic garbage bag or some other white, flat object beneath the muffler tip, extended out a few feet.
2. Start the car and warm it to full operating temperature.
3. Apply sufficient throttle to raise and lower the revs several times—between idle and 3000 rpm or so should be fine.
4. Check said white, flat object for debris. Distinguish soot from pre-cat material, if any.

Check #3: Check Your Oil
When debris has entered the combustion chamber, oil consumption will generally increase. Toyota specifications allow for the consumption of 1 quart every 1500 miles, though anything approaching that type of consumption should signal something is amiss.

Low oil doesn't necessarily mean you've ingested pre-cat material, but low levels can also kill your rod bearings, and so is a universally bad thing no matter the cause. If you find that your MR2 Spyder is consuming too much oil, pre-cat failure is simply

AW11 & ZZW30 Staged Buildups

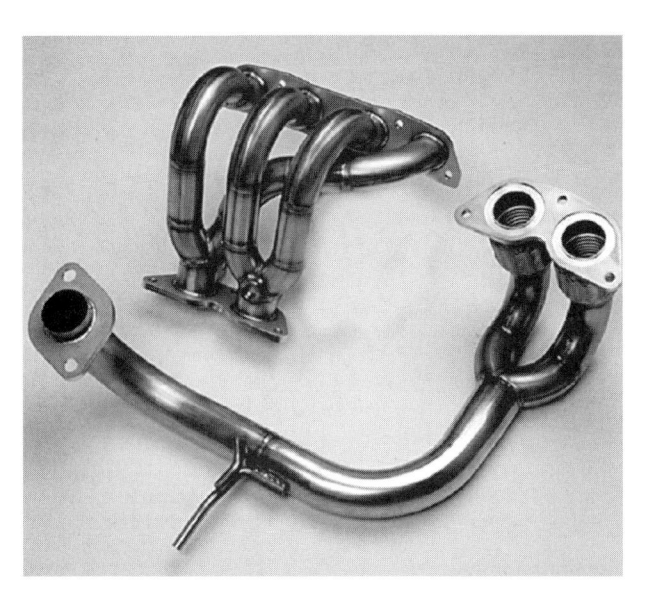

One option for dealing with potential catalytic converter failure on the ZZW30 is to simply replace them with an aftermarket exhaust manifold, like this beautifully crafted piece from Power Amuse. Photo courtesy Power Amuse.

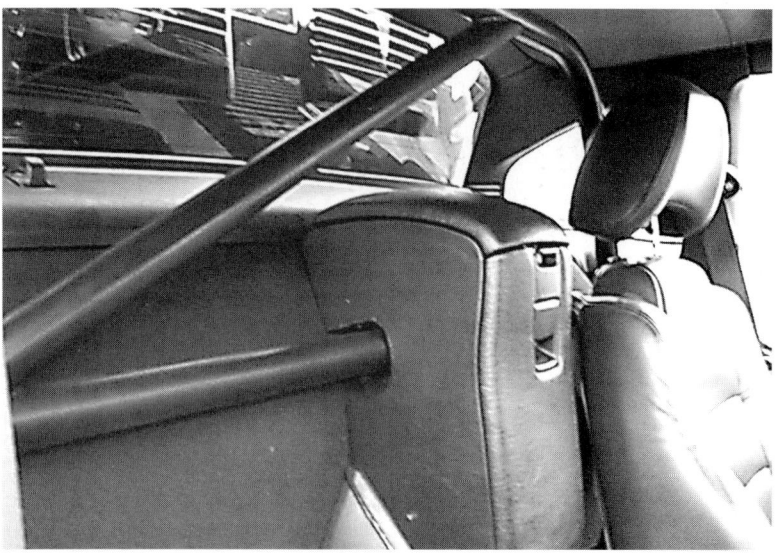

In addition to offering safety in the event of a rollover, a roll bar also helps increase chassis rigidity as well, especially crucial on convertible sports cars. A roll bar from Autopower is shown installed in an SW20.

one potential, and is probably down on the list some, too. If you are consuming a noticeable amount of oil, see your mechanic.

Correcting a Failed Catalytic Converter

If you find that you've got an issue with a failed pre-catalytic converter, one option is to gut the pre-cats—that is, pull off the manifold, take a hammer and screwdriver or chisel, and pound out the ceramic pre-cat material. The material comes out surprisingly easy, and the entire process can be finished in a few hours. The problem arises in being certain that you got all that junk out. The irony would be glaring if, in tearing 99.8% of the pre-cat material out, the 0.2% you left behind made its way into the head as soon as you started the car.

Your best bet, assuming your car isn't a daily driver in an emissions-regulated area, is to simply replace the stock exhaust manifold and pre-cat with a header. Several manufacturers make a header for the 1ZZ-FE, and though few will yield much improved performance on a stock engine, your engine can't be ruined by a failing pre-cat that's in the dumpster. The chance of your 1ZZ-FE failing is very low, so don't get in a panic—but don't ignore the potential either.

Stage I
- Roll bar
- Anti-flex plate
- Strut tower braces
- Alignment
- Header
- Brake pads, ATE Type 200 fluid

Roll Bar—In the event of a rollover, you'll be glad you have one. More important for performance, and even ride comfort, is the enhanced chassis stiffness added by the bar. Again, this is especially important in a convertible.

Of course, a simple roll bar won't stiffen things up the way an entire cage would, but a cage isn't practical for a street car anyway. Autopower has been doing roll bars for years, and there is an array of vendors knowledgeable of their product. For the majority of applications, a heavily padded Autopower roll bar, or professionally constructed custom unit, is your best bet.

This would also be the time to strongly consider a hardtop. Though these are expensive, and generally used seasonally, subjectively they can enhance the visual lines of the Spyder tremendously, and should theoretically increase rollover protection as well.

Note: Be sure to check appropriate motorsport regulations if you race your MR2—there are construction and material requirements to officially sanction a roll bar or cage. Consider factors like seat adjustment, rearward visibility, and compatibility with the hardtop as well before making your choice.

Anti-Flex Breastplate—The breastplate is instead an anti-flex modification for the MR2 Spyder. This plate is bolted to the underside of the car, basically bracing the two footwells together. It functions similarly to a roll bar, reducing chassis vibrations and enhancing the chassis's resistance to twist.

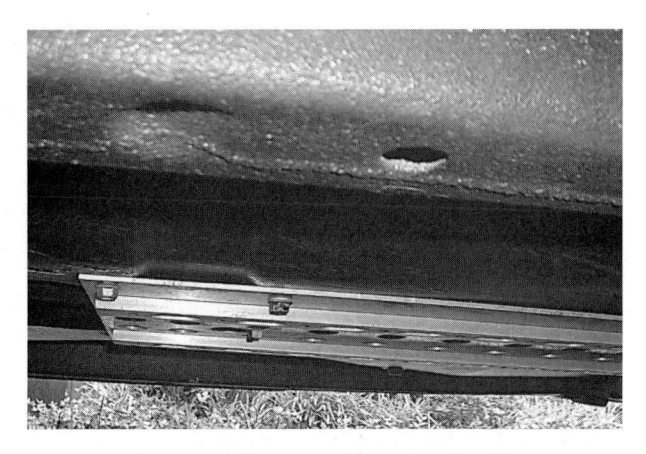

The anti-flex plate basically braces the footwells together, much like a massive strut tower bar beneath the car. This modification is especially beneficial in early ZZW30s made prior to 2003, before Toyota began stiffening things up themselves in the production models.

This modification is especially beneficial in the 2000–2003 MR2 Spyder, as Toyota began similar structural reinforcements of their own in sometime in 2003, but even on 2003+ models, it has a noticeable effect. Any car with the top chopped off, even those designed from the ground up as convertibles, can benefit from chassis bracing, especially entry-level convertibles (price-wise) like the ZZW30. Cowl shake, turn-in, and general handling precision all benefit immeasurably from a braced, stiff chassis. As a bonus, the install is a breeze.

As far as brand goes, there aren't a lot of options here. Cusco manufactures an anti-flex plate, and some industrious members of the ZZW30 community have as well; Corky's breastplate is one such product.

Strut Tower Bars—This is a modest but otherwise faultless modification that isn't brand-sensitive. Do stay away from the $20 eBay braces, though. Bars that you can bend in your hands probably won't do much to keep your chassis from twisting.

TRD sells front and rear strut tower braces, and underbody 4-point and two 2-point braces as well, the latter braces as a kit. Which pieces you end up needing depend on other chassis enhancements you've made, and any overlap that may exist between them—as in the anti-flex plate step on the previous page, for example. You may only need strut tower bars at this point, the point being that you are doing what you can to enhance the overall stiffness of your ZZW30 chassis, for benefits in performance, feel, and ride comfort.

Note: Fensport in the UK can often be a good source of both AW11 and ZZW30 parts.

TOYOTA CRASH BOLTS
Toyota Part Numbers

Front
Small: 12mm 90105–14147
Medium: 13mm 90105–14146
Large: 14mm 90105–14140

Rear
15mm 90901–05001

Some owners have used the smaller front crash bolts, with washers, in the rear as well, though I have never used that setup, so mark it down as an interesting possibility, but read with caution.

ZZW30 ALIGNMENT STARTING POINT

Camber Front: -1 to -1.5 degrees
Camber Rear: -1.25 to -1.8 degrees
Toe-in Front: 0 degrees
Toe-in Rear: 0 degrees

Note: Be sure strut bolts are have sufficient torque to help maintain alignment settings.

Alignment—You have no idea how your car really handles—and thus what to change and what to leave alone—until you have a proper alignment done. Use OEM "crash bolts," available from most Toyota dealers (see chart for part numbers), or camber bolts from Whiteline. As with the SW2x bolts, the smaller the bolt, the more physical movement/available adjustment.

Remember, experiment with alignment settings (see chart above for a baseline starting point) until your MR2 reacts how you want it to. Alignment is not a set-it-and-forget-it kind of thing, and requires attention as components wear, bumps are hit, and your use of the car changes. In fact, take advantage of every adjustable setting your suspension has until you achieve the handling balance and feel you desire—alignment, sway bar settings, shocks settings, and tire pressure. The above simply constitutes an appropriate range to begin fine tuning your alignment to suit your setup and driving style.

Note: An alignment can lose its specification, so be sure to check your alignment somewhat regularly, especially if you notice significant changes in handling character or tire wear.

Header—More as a prevention for pre-catalytic converter failure, and less for performance, but we'll

AW11 & ZZW30 Staged Buildups

The ZZW30 is extremely well-balanced and generally more forgiving than the SW2x. Adjustable shocks will help improve those virtues even further. Photo courtesy Toyota Motor Sales.

Some cars accept larger wheels better than others. The ZZW30 looks quite good with 17" wheels due to its bulbous shape and short overhangs. Photo courtesy Aaron Bown.

take what we can get output-wise on the lukewarm 1ZZ-FE—just don't expect more than 6–8 whp.

Brake Pads, Fluids—Being lighter, and having slightly larger brakes than the 200-lb. heavier AW15, and also possessing the Electronic Brake-Force Distribution system, the MR2 Spyder isn't as immediately in need of a brake upgrade as previous generations, but there is never a bad time for a quality set of pads and high quality brake fluid.

Note: Brake pads and ATE fluid really are obviously not critical for a commuting ZZW30, but of course we're assuming you and your mid-engine, 2200-lb. Spyder will be doing more than commuting. With the above setup, brakes should remain fade-free under all but the most punishing driving conditions.

Stage II
- Lightweight battery
- TRD Sportivo handling kit
- Front and rear lower braces
- B&M short-shifter
- 2000–2002 ZZW30: 2003–2005 OEM wheels

Lightweight Battery—The benefit here is easy to guess at: lower weight. Note that this modification will be less desirable in colder climates where cold cranking amps on cold mornings might outweigh the benefit of a slightly lower weight. Figure on saving 10–15 lbs. here.

TRD Sportivo Handling Kit—TRD offers the Sportivo kit which consists of shocks, springs and sway bars. The kit is a handling package that represents a Stage II level of handling. This kit is a matched set of shocks/struts springs, and sway bars with rates designed and engineered to complement one another. Surpassing the performance offered by this kit would be difficult to match. The Sportivo kit from TRD is a solid modification, and will vastly increase the lateral grip, responsiveness and overall handling ability of your MR2 Spyder, without risk or significant compromise.

An alternative package if you want to go it alone would be Koni shocks, Whiteline adjustable sway bars (22mm front, 18mm rear) and stock springs.

Front Lower Brace—Similar to the breastplate, and strut tower bars, lower tie bars "tie together" the MR2 Spyder suspension, providing the stated benefits of a stiffer chassis. A lower Cusco front bar will do exactly that, and will further the work we've done already to stiffen up our open-topped roadster's chassis. Brand isn't critical, either—braces are braces provided the metalwork quality is up to par—so you can save yourself some money here if you can go with a custom bar through a knowledgeable tuning shop.

Bear in mind that as the chassis is stiffened by modifications, future modifications will have less effect; there is simply less room for improvement. Lower tie-bars and braces are a big step in the right direction and as such the need for this continued chassis stiffening should be consistently reevaluated based on how the car feels, what kind of power modifications your MR2 Spyder has, if the car has ever been wrecked and so on. There comes a point, however, where the chassis will be stiff enough that further improvements will be less dramatic.

Short-Shifter—To reduce our shifter throw and tighten up the driver-to-car dynamic, we're looking for a short-shift kit here. The short-shift kit from B&M will shorten your shifter throw by 38%.

The ZZW30 is such a well-balanced performer stock that modifications can focus on safety, durability and even visual components. This MR2 Spyder features a hardtop, Volk wheels, Recaro seats and a carbon fiber hood to reduce front-end weight. Photo courtesy Aaron Bown.

Stage III
- Eliminate wheel hop
- ACT clutch
- Fidanza flywheel
- Quaife LSD
- KW Variant 3 coil-overs
- Forced-induction or engine swap

Addressing Wheel Hop—Wheel hop occurs when spinning, contorting tires transmit unwanted force throughout the suspension and can't hook up. In other words, you can't get the power to the ground. It is basically a battle of physics between the spring, the tire and the road, all ricocheting force back and forth until the torque applied to the spinning wheel changes, gets traction, or something breaks. This is a problem traditional American muscle car owners have been dealing with for years, but as far as MR2s go, is mainly contained to the ZZW30. Weight, suspension design and torque all play a role in wheel hop, and there are a variety of ways one can address it.

Wheel hop can be a troublesome issue at any stage of development, but it is considered a Stage III modification because most people don't launch their ZZW30 hard enough until they increase performance to this level. Certainly it can be damaging at any level of performance, and can break axles even on a stock Spyder, but until now the best advice is to simply drive around it—or don't do what causes it, if that isn't too simple a remedy.

Wheel hop can occur under a variety of circumstances, but is generally associated with very aggressive braking and launching maneuvers.

Seeing as how the ZZW30 has ABS, wheel hop while braking shouldn't be an issue—but launching is another story. The best solution is to stiffen things up. Engine mounts, shocks, springs, suspension bushings, strut tops and other relevant chassis bracing. The less initial compliance your car sees, the less violent the oscillation back will be, and so on. Effectively combating wheel hop, like most modifications, can't be narrowed down to a single brace or retorqueing of a bolt, but the above changes, most of which have been performed by Stage III anyway, will go a long way in preventing wheel hop from rearing its shuddering, ugly head.

Note: Solid strut tops/pillow mounts like those available in many coil-over setups, radically diminish ride quality, but hard rubber or custom urethane strut tops can be an excellent trade in ride quality and performance when you're looking to eliminate compliance in your suspension setup.

SWAPPING 2000–2002 ZZW30 FOR 2003–2005 OEM WHEELS
ZZW30 OEM Wheel And Tire Sizes

2000–2002
Front: 185/55/15, 15x6
Rear: 205/50/15 15x6.5

2003–2005
Front: 185/55/15, 15x6
Rear: 215/45/16, 16x7

Some cars accept larger wheels better than others. The Honda CRX, for example, looks ridiculous with any wheel larger than 15", and the AW11 doesn't accept large wheels well, either. The ZZW30, on the other hand, looks quite good with 17" wheels due to its bulbous shape and short, interesting overhangs. Be mindful, though, of any increase on unsprung weight. Here you'd be best served by wide, lightweight wheels, and as mentioned, the 2003–2005 16x7 ZZW30 is not a bad place to start if you've got an early Spyder.

An alternative to the 2003–2005 OEM wheels is the Enkei RPF1. This is an extremely lightweight wheel at only 9.5 lbs. per wheel, saving 5.5 lbs. over the stock wheels, which are closer to 15 lbs. each. They are also very inexpensive like the Kosei K1, which allows you to spend money saved on the parts that actually make your car faster. This would allow you to purchase aggressive R-compounds for one set of wheels—the Enkei RPF1s I'd guess—and use the other for daily driving. Wheel taste is very subjective though. The gist here is lightweight wheels with a wider contact patch.

Note: Be aware that altering wheel sizes from stock, gearing and speedometer operation can be affected. To avoid this, care must be taken to try to maintain the stock aspect ratio.

AW11 & ZZW30 Staged Buildups

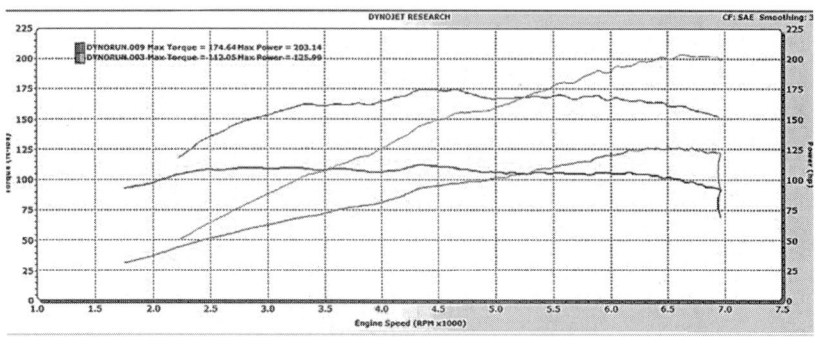

The Hass Pro Turbo kit at 6 psi, as setup on the car being dyno-tested here offered gains of 76 hp, and by 3000 rpm, an increase of 50 lbs-ft. of torque. See Chapter 4 for additional photos of the kit. Image courtesy Sherman/Haas.

On the KW Variant series coil-over, the coil-over housing itself is made of stainless steel and the adjustable collar from anodized aluminum. Photo courtesy KW Automotive.

The reduced deflection of firmer urethane bushings can help reduce the incidence of wheel hop in the ZZW30. Simply modifying your launching technique can help as well.

Clutch/Lightweight Flywheel/Limited Slip Differential (LSD)—The ZZW30 stock flywheel weighs about 15.5 lbs.—fairly heavy for a zippy roadster. The benefits of a lightweight flywheel in any MR2 aren't significant enough to pull a perfectly good clutch just to install one, but if everything is out anyway, look at a Fidanza flywheel.

Same with an LSD. The MR2 transaxle, in every generation, comes apart in similar ways: generally the engine is at least detached at all but a single engine mount, and the transmission "tilted" down, allowing the transmission to be pulled off—at which point, the differential, clutch assembly, and flywheel are all within reach. Because of the vast savings in labor—either time, if you're doing it yourself, or money if you're paying someone else to do it—this can be considered a package modification.

Technically a Genuine Toyota clutch is plenty, though if you're going to pull everything for a lightweight flywheel and/or LSD install, you can consider a stouter (but still organic) clutch from ACT for future power potential, at your discretion of course.

A Quaife ATB LSD, part QDF21E (the Quaife LSD also features a lifetime warranty) is the best choice for an LSD. If you're wanting a a clutch-type LSD, a 1.5 LSD from Kaaz is available as well.

Coil-Over Suspension—If you really want to continue the development of your ZZW30 chassis, coil-overs are your next step. KW Variant 3 lightweight coil-over suspension offers a nice upgrade over even the TRD Sportivo kit, adding ride-height adjustability, corner-weighting potential and durable stainless steel construction. Which coil-over suspension you choose, as we've stated elsewhere, buy from a knowledgeable vendor, preferably in the United States.

Forced-Induction or Engine Swap—The ZZW30 is certainly quicker than an AW15 in a straight line, and probably the AW16 too, which has never been accused of being slow. Still, those "slower" normally aspirated MR2s cost about $20,000 less than a newer MR2 Spyder. Regardless of a decent—especially for a 2200-lb. car—138 horsepower starting point, the dreadfully tame 1ZZ-FE has very little seat-of-the-pants power. The mass market's main MR2 Spyder competition was the Miata, and it was quicker than the Mazda, but

113

Toyota MR2 Performance

A version of this turbocharger, the GT28RS, is used on the ZZW30 Hass Pro Turbo kit. This turbo has proven well-sized to the 1.8l-2.5l engine size. Photo courtesy Steve Guttman.

it is safe to say that the MR2 Spyder's straight-line performance leaves something to be desired, and doesn't do justice to the extremely well-balanced and able chassis—especially one upgraded with Stages I and II. In this case, a 1ZZ-FE forced-induction kit, or an altogether new engine, could prove useful.

One could certainly seek out the traditional breathing enhancements to increase the horsepower of the 1ZZ-FE in their MR2 Spyder, but as with 4A-GE bolt-ons, the bang-for-the-buck just isn't there, not to mention that type of aftermarket support. Instead, there are both bolt-on supercharger and turbocharger kits available for the MR2 Spyder, as well as the 180 hp 2ZZ-GE from the Celica GT-S, and the K20A from the Acura RSX Type S.

A Hass Pro Turbo kit is probably the most common choice for ZZW30 forced-induction in the United States, and is popular for a reason. If it were my ZZW30, I'd go with a 2ZZ-GE or K20A2 engine swap for long-term durability and ease of running, but if one absolutely had to go turbo, the Hass Pro Turbo kit is the most well-sorted and grassroots-proven.

Hass Pro Turbo Spyder Turbo Kit
Performance Gain: Up to 275 whp

There are other available kits on the market, like the APEXi, for the MR2 Spyder, and there are sure to be others popping up from time to time, but the years of experience developing turbocharger kits by Hass separates their kit from larger manufacturers. The dual ball-bearing design not only increases the spool of the turbocharger, but its durability as well, a factor not to be ignored. The inclusion of an engine management system from GReddy, the piggyback e-Manage, is a nice feature, but only as helpful as the person tuning with it. A cheap alternative to a 2ZZ-GGE swap, though turbocharging an engine produced by the factory as normally aspirated is always a calculated risk.

Specifications:
- Garrett GT28R dual BB water-cooled, dual ball-bearing turbo. Cold side: 60mm 60 trim compressor wheel in a 0.60 A/R housing. Hot side: 53mm, 62 trim turbine wheel in a 0.64 A/R 5-bolt discharge housing (turbine housing thermal coating optional). Internal wastegate.
- Stainless steel thick-walled exhaust manifold, (1/8") 1 5/8" ID tubing with 1/2"-thick flange. High-temp aluminum ceramic thermal coating optional.
- 3" mandrel-bent downpipe with 5-bolt discharge flange and smooth-merge wastegate return.
- 3" high-flow Cat (test-pipe optional and is a 4" round 12" long resonator for no loss in sound control, a feature not found anywhere).
- 3" mandrel-bent aluminized single exit exhaust with straight through muffler design. (Dual outlet exhaust optional with resonance tuned chamber for unique sound.)
- Custom machined OEM pencil-style injectors to work with the factory Toyota fuel rail and provide extra fuel; no wire "cuts" or "splices." Toyota runs a higher factory fuel-pressure which extends the range of these injectors to flow 410cc.
- Titanium turbo water inlet/outlet fittings with aluminum banjo bolts and crush washers. 1/2" stainless steel coolant feed and return lines with couplers to tie into factory coolant line.
- 4' stainless steel braided oil feed line with compression crimped -4 AN female fittings. Stainless steel reverse flare inlet fitting with -4 AN male end feed fitting. Brass distribution tee for oil pressure sending unit and inlet line. 1/2" stainless steel braided oil return line. Aluminum turbo oil drain flange with threaded 3/8" NPT center hole. Two brass 3/8 NPT to 1/2 hose barb fittings for oil pan and turbo oil drain outlet.
- 2" mandrel-bent, lightweight intercooler piping. Pipes are coated with aerospace standard gunmetal gray metallic epoxy finish. Inside of piping is aluminized to prevent corrosion.
- The IC core is a bar and plate Blackstone high-density core unit and measures 6.25"x18"x3.5" (not including end tanks). It works as a fairly large heat sink so it can absorb a fair amount of heat giving the car a "stored" cold air charge until vehicle speed resumes. The IC has cast aluminum end tanks just

AW11 & ZZW30 Staged Buildups

like the "big" companies use such as GReddy, HKS etc., and not welded sheet metal with harsh angles with cast aluminum elbows. The intercooler core when tested at an air inlet temp of 250°F at 14.7 psi, with a cooling air temp of 75°F lowered the discharge temperature to 124°F. Includes mounting hardware/brackets and air deflector preinstalled on unit. Stage II only.
- Four-ply Nomex couplers for Intercooler pipes, TB connection and turbo inlet pipe with one 2"–2.5" reducer for throttle body outlet.
- 1.5" x 7.5" self-adhesive thermal wrap for compressor outlet reducing coupler and oil feed line (extra heat protection to keep long term integrity of coupler and line).
- T-Bolt clamps for couplers, reducer and turbo inlet pipe.
- Factory Porsche 911 Twin Turbo Blow-off Valve recirculated into turbo inlet pipe.
- High-flow 3" turbo inlet tube with conical K&N air filter. 2.5" MAF tube in a blow-through configuration calibrated to kit specifications, allowing venting of the recirculation valve to atmosphere for better drivability and accommodation of the SMT (transmission) models.
- Reducer for factory vacuum line to 3/16" vacuum line and brass tee fitting for blow off valve and wastegate actuator. 4' of standard rubber vacuum line.
- 30 psi boost gauge.
- GReddy E-manage with injector and timing harness preprogrammed and ready for install. Dyno-proven base tuned maps included for ignition and fuel for 91 octane and 11.5–12.0:1 A/F ratio up to 270 whp.
- Other: Mobil 1 or K&N oil filter, one range colder copper plugs, tube of RTV for oil pan sealing, zip ties, electrical connectors, etc., all included for stress-free installation.

Beyond

As with the AW11 and SW20, removing weight is always a possibility for further performance enhancement, the wonder of it that it will help your MR2 do everything better—go, stop, and turn. Though a 2200 lb., 120 hp ZZW30 won't punish

The ZZW30 was built as a lightweight sports car. Tadashi Nakagawa, Toyota's ZZW30 chief engineer, is on record saying that handling was a top priority for the ZZW30, something confirmed with the choice of the lukewarm 1ZZ-FE engine. Still, if you're willing to pay the piper, a turbocharged 1ZZ-FE is possible. Photo courtesy Davey Kerr.

brakes like a 2800 lb., 340 whp SW20, brake fade is dangerous, and can end a Saturday of hot-lapping frustratingly early.

Aerodynamics are also an underappreciated sector of performance that OEMs like Toyota are able to generate significant results with. Development costs and resources here are extremely high though, generally beyond the reach of the average owner/tuner, but if you're looking to fully optimize what you've got, there is some fertile ground here.

The most obvious direction to continue a ZZW30 development though, beyond continuing to tweak and readjust what you've already done, is towards additional horsepower. Since the ZZW30's production is much more recent than the AW11 and SW2x, there is still fairly active development in aftermarket parts, including some interesting supercharger forced-induction options for the 1ZZ-FE engine. Your local ZZW30 online communities will be your best source for current information on newly developed parts, but remember—being a pioneer isn't all glory.

K20A2 ZZW30 ENGINE SWAP CONSIDERATIONS

It could be that the available turbocharger or supercharger kits on the market aren't your thing. In fact, you don't like the 1ZZ-FE at all, and want to be done with it. When swapping a new engine into the MR2 Spyder, most folks reach for the 2ZZ-GE from the Celica GT-S, the powerplant many owners felt should've been in the car all along. Rcrew Racing thought different, and cribbed a K20A2 from an Acura RSX Type S/2006+ Honda Civic Si. Their careful installation and custom-manufacture process follows.

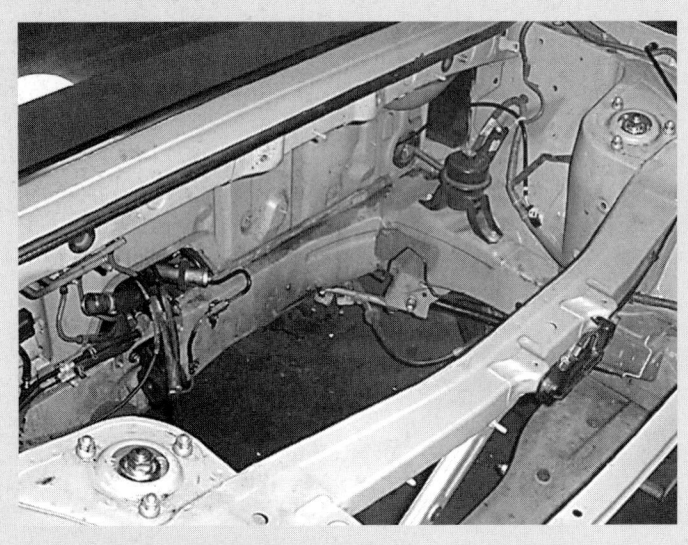

Because of the mid-engine configuration, transmission cables, coolant lines, and other various lines must be routed through the firewall. Photo courtesy Rcrew Racing.

In swaps like these, many custom pieces have to be fabricated. This mounting bracket is for the transmission cables. Photo courtesy Rcrew Racing.

Custom engine mounts are necessary, too. Not only must these be fit and precisely located, but they must also be designed strongly enough to withstand the significantly increased output most swaps produce. Appropriate engine mounts are the foundation of any engine swap. Photo courtesy Rcrew Racing.

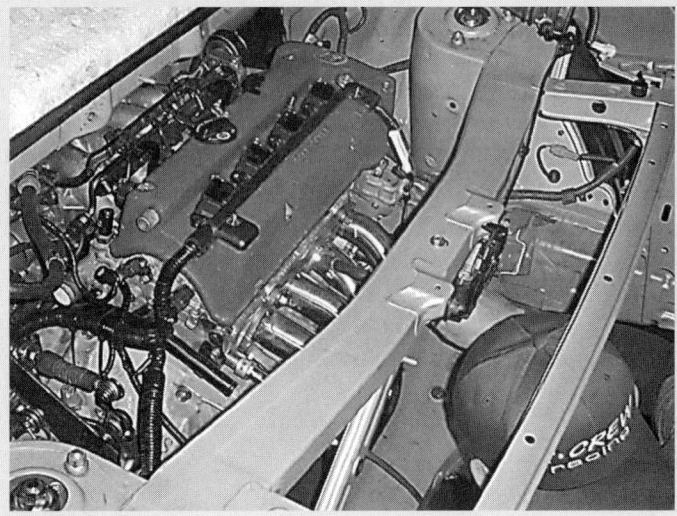

Putting the finishing touches on the exhaust manifold. The exhaust manifold clearance is not nearly as tight as the intake manifold clearance is on the firewall. Photo courtesy Rcrew Racing.

AW11 & ZZW30 STAGED BUILDUPS

The exhaust manifold clears the rear crossmember easily and simply needs a mid-pipe and a muffler to be finished. Photo courtesy Rcrew Racing.

From this angle you can see the landscape of the engine bay, with the lightweight battery to the left, and the easy-access-to-oil-changes oil filter to the right. So far, the swap looks factory-clean. Photo Rcrew Racing.

The full exhaust system, with a muffler by Borla. This custom exhaust system doesn't use a flex pipe; some systems, like the (SW20) GReddy Power Evo, simply depend on the exhaust hangers to allow movement by the exhaust system. Photo courtesy Rcrew Racing.

Chapter 6
SW2X Staged Buildups

The SW20 is an aggressive looking, potent platform that perfectly embodies the spirit of the modern sports car. It also responds extremely well to the right modifications. Photo courtesy Brad Armstrong.

1991–1995 SW2X
SW2x Performance Strengths
- Handling potential
- Braking performance
- Aerodynamics
- Engine output and potential (3S-GTE)

SW2x Areas Needing Improvement
- Handling stability
- (1991–1992 SW20) Transmission performance
- Powerband shape

SW2X STOCK ASSESSMENT

At around 2800 lbs., the SW20 is not a light car, but it handles its weight well. A stock example will produce 0–60 times in about 6 seconds, passing the quarter-mile in the mid-to-high 14-second range.

The measurement of its stock handling was always superb, with major automotive media publications consistently reporting 0.90g's of lateral grip on the skidpad, and slalom times matching or eclipsing those from many higher-priced Porsche and BMW models. However, the SW20 can be a handful if not driven with precise input, and our suggested modifications will focus on making the handling more predictable and adjustable enough to suit a variety of driving styles and conditions, while pushing the handling performance envelope. The stock braking performance is world-class, even by today's standards, though some of that comes courtesy of the mid-engine configuration. And even the best factory brakes can benefit from a few enhancements,

Autocross competitions bring out the best in the SW2x due to its natural ability to launch hard bestowed by its mid-engine, rear-drive configuration. Photo courtesy Hsun Chen.

namely in an effort to enhance fade-resistance.

The interior is borderline perfect, though it could use a few more gauges and electronic devices to provide more information on how the engine is performing during aggressive driving.

The engine makes 200 horsepower, which is far from disappointing, but starts falling off noticeably past 6000 rpm. This is perhaps the biggest criticism of the stock MR2 Turbo. The CT-26 turbocharger has a turbine housing that severely restricts airflow at higher rpm, and while it is a

SW2X STAGED BUILDUPS

Braking enhancements are mainly beneficial on the racetrack—commuting will not fade your stock brakes. But if tracking canyon runs and back-country roads are your thing, they are a worthy upgrade. Photo courtesy Vespremi/Crites/Smith.

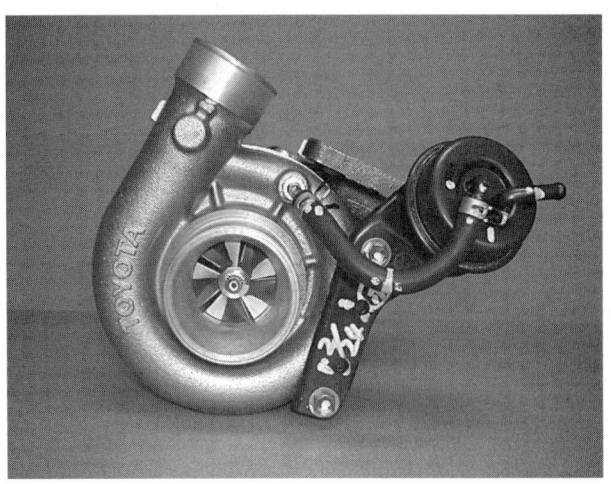

If we're going to champion the strong points, we've got to be honest about what's not strong. The stock CT26 turbocharger is one of those not-strong points. Enter the CT20B, Toyota's replacement for the CT26. Photo courtesy Brian Hill.

The SW2x interior is clearly one of its strongest points. Once you start radically increasing power production though, you need more information than the poor stock boost and center-weighted temperature gauges. You need real data. Photo courtesy Vespremi/Crites/Smith.

One of the most well-balanced and carefully planned buildups of the SW2x in the United States is David Vespremi's MR2, which started life as an SW21.

moderately capable turbo for normal street driving, a larger turbo with more airflow is a better choice for high performance applications.

The SW20 has the most accessible platform for increasing performance, with more options available than either the AW11 or ZZW30. For the SW20, engine swaps or other drastic modifications are not necessary for dramatic increases in performance, and while paling in comparison to current performance darlings like the Subaru WRX, Mitsubishi Evo, and Nissan 350Z, the SW20 still enjoys enough aftermarket support that tuning projects don't have to turn into a sequence of custom, one-off chores.

In summation, as with the AW11 and ZZW30, in the SW2x we are looking for brakes with increased fade-resistance, an improved powerband, adjustable suspension components, and more predictable handling.

This guide assumes, like the vast majority of SW22s, you have a Gen II 3S-GTE. As mentioned, the majority of these changes will be effective across the generation of 3S-GTEs, and turbocharged cars in general, but many are Gen II 3S-GTE-specific,

SW21: A SPORTS CAR WITH A CAMRY ENGINE

The SW21 is a unique vehicle. Possessing a mid-engine design, sexy, almost pitch-perfect lines, and a sturdy-but-boring engine from a Camry, it is automotive melancholy. Like all MR2s it handles particularly well—in fact, among every variation of MR2 produced, the SW21 and ZZW30 are the best handling, improving specifically upon the handling of the more-celebrated SW20 with a lighter weight, and a less-temperamental delivery of power more conducive to limit-handling. It is also every bit as durable as the SW20, if not more so, and less expensive to run year in and year out.

It also produces underwhelming acceleration. How much this bothers you depends on your expectations, but this makes it a quite different car to drive, own, and modify. Even though the SW20 and SW21 share the same body style and suspension design, they offer quite different driving experiences, and so a separate SW21 staged buildup would focus almost entirely on handling and braking—basically the buildup you see below, minus the horsepower stuff.

Unlike the 4A-GE, the normally aspirated 5S-FE does not like to rev, and its relative lack of responsiveness and upper rpm performance make significant engine modifications a waste of time for most owners. I'm sure there are owners out there who would heartily disagree with that last sentence, but it is the simply the experience I've had. Spending thousands of dollars modifying a 5S-FE is missing the point.

Due to the similarity of the SW20 in overall setup and design, and the lack of aftermarket support for the 5S-FE engine in the SW21, this guide focuses on the SW20, though similarities can be obviously deduced where appropriate—suspension and alignment, for example.

Jeff Fazio's SW20 has been methodically built, tuned, raced, re-tuned, rebuilt and raced again, producing an MR2 that has no glaring weaknesses, all done on a budget that might surprise you. Photo courtesy Jeff Fazio.

the Gen III 3S-GTE turbocharger not as critical as it is on the Gen II, for example.

STAGE I

- Alignment
- Fresh Genuine Toyota ignition components
- Boost gauge
- Exhaust
- Tokico Illumina adjustable shocks
- Redline MT-90 transmission fluid
- ATE Type 200 brake fluid, Porterfield R4S pads (or equivalent), Speed Bleeders
- Hella H4 bulbs
- 1993–1995 OEM wheels

Alignment

As discussed in Chapter 3, the ideal alignment can not only improve handling stability, but can improve turn-in, road-holding and overall feel. One way to achieve ideal alignment specifications is through the use of crash bolts.

A crash bolt is so called because they are often used by repair shops to gain satisfactory alignment settings on a car that has been crashed, and thus out of alignment. These offset bolts are used in the strut housing, and due to their unique size allow for more ranged alignment settings. These are OEM pieces available for purchase through Toyota dealers, available as three grades of crash bolt—short, medium and long. We're after the smallest ones, and we'll need four—two per side, Toyota part number 90105-15006. The appropriate nut is Toyota part number 90179-15001.

Different Applications, Different Alignment Needs—Keep in mind what you're doing with your MR2 before you specify to a shop what alignment you're looking for. While negative camber increases grip when the car turns in, too much of that camber means that when the suspension is not loaded, the tires are sitting at an angle not ideal for the hard launches a drag strip would see, not to mention straight tracking down the road while commuting and running about.

Negative camber will also wear the tires more quickly, while adjusting the caster and toe can make the car track oddly, making for a commute that requires constant correction at the steering wheel. Even your brand of tire, its pressure, and sidewall height and stiffness can have an effect on which alignment is ideal for your MR2. Some performance tires, such as the BF Goodrich Comp T/A R1, feature sidewalls that allow for dynamic changes in negative camber as the sidewall flexes in predetermined, beneficial ways. On this tire, less

SW2X STAGED BUILDUPS

Alignment becomes absolutely critical during at-limit cornering, allowing the contact patch to do its best work by remaining ideally positioned with the road under a variety of suspension-load conditions. Stability under fast transitions, like a slalom on an autocross course, is directly impacted by wheel alignment. Photo courtesy Hsun Chen.

SW2X ALIGNMENT BASELINE SPECIFICATIONS

- Camber Front: -2 degree
- Camber Rear: -1.5"
- Caster: +6
- Toe-out Front: 0" to 1/8" per side
- Toe-in Rear: 3/16"

Note: 1993–1995 models can start with a little less rear toe-in, 1/8" as a starting point.

Note: Caster is not adjustable on 1993+ MR2s, one of the very few disadvantages they have over the early 1991–1992 models.

Remember, experiment with alignment settings until your MR2 reacts how you want it to. Take advantage of every adjustable setting your suspension has until you achieve the handling balance and feel you desire—alignment, sway bar settings, shocks settings, and tire pressure. Rather than leaning on one setting—huge sway bar or huge amounts of negative camber, for example—consider the handling of your MR2 as a result of system, not a single part.

negative camber would then be necessary.

I recommend that you start with the specifications in the chart above as a baseline for alignment. From these settings you can experiment—adjusting rear toe or front camber, for example, to produce the desired handling tendency and tire wear you can live with.

New OEM Ignition Components

If you haven't already redone your ignition components, do so at this point before continuing. New Genuine Toyota distributor cap, rotor and plug wires, and an OEM igniter. Spark plugs are the critical to your MR2's maintenance. Run the right plugs for your setup, and change them often.

Which plugs and how often depends on what heat range is best suited to your particular setup—how much boost are you running? A good all-around choice would the NGK 6097 gapped to 0.028." It is one heat range colder than the stock plug Toyota recommends, offering better detonation resistance.

Note: Check your base timing, see that it is set between 8–10 degrees, and leave it there!

Aftermarket Boost Gauge

The factory boost gauge is barely worth keeping. It lacks resolution, response and accuracy, and if you're running more than stock boost, you'd be better served by the voltmeter Toyota supplied the

OEM ignition components are all that are necessary at this power level, and well beyond in fact. These TRD of Japan-specification 3S-GTE plug wires can be considered OEM and can add a little visual pop to the engine bay. A less expensive alternative is to simply use Genuine Toyota 3S-GTE wires. Photo courtesy Brian Hill.

121

The factory boost gauge is really more of a vague detector of vacuum than a boost gauge. It lacks resolution and response, and needs to be replaced before you turn up the boost. This gauge from APEXi is graduated in easy-to-read psi. Photo courtesy Vespremi/Crites/Smith.

The B-pipe is the pipe between the primary cat and the muffler itself. It contains a secondary catalytic converter that represents a significant restriction in the exhaust path. Photo courtesy Jeff Fazio.

SW21 with in its place. We're replacing the stock gauge now, before we turn up the wick on our very potent 3S-GTE. A nice choice would be an electronic boost gauge with peak hold (it captures the highest reading until reset). Manufacturers like APEXi, HKS, Blitz, Omori, VDO and Auto Meter make excellent gauges. The point is to have a real boost gauge to tell us what's going on. There are also many engine management systems, multi-meter systems, and even fancy turbo-timers that can monitor boost as well, and the digital nature of many of these pieces makes using the peak-hold function easier, a good thing. All-in-one turbo-timers and boost controllers like these can save clutter in the interior, and money as well. Personally, I prefer an A-pillar dual gauge pod holder to house the gauges, the driver-side A-pillar off to the left making it unnecessary to take your eyes (too far) off the road.

Custom installs on top of the steering wheel, in place of the factory boost gauge, and in DIN panels under the CD player or over the ashtray are available as well, each with their own respective pros and cons.

Exhaust Modification

This step is purposely vague. What it boils down to is either: A) modifying your entire exhaust system right now, from the downpipe to the muffler's tip and everything in between, or B) removing bottlenecks (like the primary and secondary catalytic converters) individually, based on budget and the level of boost you're running.

At minimum, the B-pipe needs to go at this stage, and your best bet is to find a reputable fabricator like KO Racing or AutoLab, or a local quality

The stock SW20 exhaust is routed through a primary catalytic converter bolted directly to the turbocharger, through a second cat just downstream into a very heavy muffler, then exits through two muffler tips in the style shown here. Photo courtesy Brad Armstrong.

muffler shape to make that happen. The B-pipe is very restrictive, and it can be replaced while still retaining the factory muffler if you're on a budget and modifying in stages. Some legal emissions-level requirements can still be satisfied via the primary cat, depending on the specifics of the smog-check in your county.

If money isn't an issue, there are literally dozens of complete aftermarket exhaust systems available for the MR2 Turbo, disproportionate to the typical aftermarket support it usually enjoys. Due to its rear bumper design, MR2s display the exhaust well, showing off careful welds and shiny steel in a very pleasing way. The Blitz Nur exhaust, and the GReddy Power Evo are good options, but exhaust systems tend to come in and out of production

SW2X STAGED BUILDUPS

Most aftermarket exhaust systems are cat-back, meaning that they bolt to a primary or secondary catalytic converter, and replace everything after. This Blitz Nur-spec exhaust bolts to either the primary cat or downpipe, whichever you've got.

The GReddy Power Evo is a wonderful performance exhaust—a lightweight, simple straight-through design. For the daily driven SW20, it doesn't work as well, lacking a flex pipe and being very loud.

The Blitz SW20 exhaust is the exact opposite of straight-through, making some funny twists and turns. The removeable silencer in the tip costs some horsepower, but significantly quiets the exhaust.

rather quickly, and are generally very expensive. General guidelines include finding one with mandrel bends, a 2.5–3" diameter, stainless steel materials and a flex pipe.

Adjustable Shocks

Koni Yellows and Bilstein shocks are high quality, excellent options, but they cost almost as much as a set of coil-overs. Tokico Illuminas arguably represent the best bang for the buck, and are available at many discount retailers as well. Their 5-way adjustability is easy to use, with preset "clicks;" the Konis offer a wider range of precise adjustments, but you actually have to count turns, which can get confusing.

Note: If you want to be competitive in an SCCA stock class, go for the Konis. Because of rules and classification, you can't go beyond adjustable shocks and a front sway bar anyway, so go for the best shock money can buy.

REPLACING SHOCKS AND SPRINGS

If you're autocrossing in a stock class, one of the few changes that you're allowed to make is the addition of aftermarket shocks. These shocks are typically stiffer on both compression and rebound, and are available adjustable as well. Shock removal and replacement is also well within the reach of the average shade-tree mechanic. The worst part is compressing the spring, which may not even be necessary depending on the spring and shock combination you are using.

Before You Begin

We'll be installing Eibach's Pro Kit progressive springs and Tokico Illumina 5-way adjustable strut cartridges on a 1993 SW20. You should rely heavily on the BGB Toyota factory manual, and other owner installs available online to accomplish this. The BGB is missing several significant helpful tips, but does offer factory torque settings, which is very useful. This is a major undertaking, at least in time required, but not an overly complex one. Don't rush it!

Make sure you know how to use a coil-spring compressor, and follow safety guidelines, as these compressors can be deadly. I strongly recommend using six-point sockets and box wrenches on all nuts and bolts heads. Many of these fasteners will likely not have been removed in a long time, and are likely impacted with rust and corrosion.

Toyota MR2 Performance

Tools Required
- Floor jack (larger is better) or lift
- Jack stands
- Wheel chocks
- 1/2" drive metric socket set with extensions and breaker bar
- 3/8" drive metric socket set with extensions
- Metric combination wrench set
- 5mm Allen wrench
- Torque wrench capable of 20-130 lbs-ft.
- Various sizes of Philips and flat screwdrivers
- MacPherson strut spring compressors
- Large (5" minimum) machinist's vise on sturdy workbench
- Long needle-nose pliers with bent tips
- Lightweight general purpose oil
- Hacksaw
- Long cable ties
- Thick bath towel
- Grinder with wire wheel
- Dremel Moto-tool with abrasive wheel
- Cleaning solvent (carburetor cleaner or general degreaser)
- Oil pan (to catch waste oil)
- Chalk or marker

Procedure

1. We'll cover the rear install, which is mechanically similar to the front. Place wheel chocks in front and back of the front wheels. Loosen the wheel nuts on both rear wheels, then release the parking brake. I jacked up the car in the center of the rear crossmember, and set jackstands at the jacking points of the door sills.

2. Remove the wheel from one side. Mark one of the holes where the stud protrudes to indicate which ones to line up afterward. There can be issues during reinstallation, and changing the orientation of the rotor on the lugs can help. It's helpful to screw one of the wheel nuts back on to keep the rotor in place. Since the rotor is only held in place by the wheel nuts and the brake caliper, it might tend to fall off later when you remove the caliper.

3. The BGB says you need to disconnect the brake line, but there's a way around that. You'll need to cut a slot about 3/8" wide in the hose bracket on the strut body. First, remove the U-shaped clip from the underside of the bracket to free the hose joint. A hacksaw works well to make the incision. You can use a Dremel with a grinding wheel attachment to smooth the edges of the cut afterward, which will help it look less "hacked," and more natural.

The brake line will pull away from the strut body after removing the U-shaped clip that anchors it.

4. Next, you'll need to remove the brake caliper, something the BGB neglects to mention. There are two 17mm caliper bolts on the back side of the hub. You can use a long breaker bar to loosen these bolts, or another method you prefer.

Support the weight of the caliper somehow before completing detaching it to avoid stressing the brake hose. A zip-tie works well here.

5. After setting the caliper aside, you'll need to remove the upper mounting bolt on the stabilizing link. I recommend strongly that you soak the nut with penetrating oil prior to the removal. It's quite easy to damage the stud if you don't. Fit a 14mm box wrench on the nut. I recommend a 6-point wrench. The stud has a 5mm hexagonal hole in its shaft to enable fitting an Allen wrench to prevent it from turning with the nut.

6. It's common for the nut to loosen a bit, then jam up as you try to remove it entirely. If you strip out the internal hole, you can get a grip on the stud from behind the bracket with a pair of thin-nose pliers or vise grips. Be very careful, though, as you

SW2X Staged Buildups

Ensure that the Allen wrench is fully and cleanly inserted into the hex hole to keep from stripping it. Suspension components see a lot of grime, dirt and corrosion.

could also ruin a stud this way. If the bolt is reluctant to loosen, try removing the lower bolt instead. You can remove the top one later after the strut has been removed. Once the nut is removed, test to see if there is any tension or compression on the link. If not, you might be able to push the stud back through the hole in the bracket, and twist it out.

Note: If the stud won't clear the bracket, or if there's too much tension/compression on the link, leave it alone until you are ready to remove the strut later on. Before you proceed to actual strut removal, find a nice, thick towel and fit it between the

After detaching the end-link, cleaning the mating surfaces and checking the condition of the rubber end is a good idea.

bottom of the strut and the axle below. This is to prevent (hopefully) any damage to the axle and boot when the strut is unbolted. My rear struts were still under some pressure after the bolts were loosened, although some users have reported otherwise. It's best to be safe.

7. Now it's time to loosen the 19mm bolts attaching the strut to the hub. Again, a long breaker bar will be necessary. Once you've removed the nuts, DO NOT remove the bolts.

Using a breaker bar requires leverage and often interesting angles with the bolts. Considering the amount of force and movement in order here, be sure the car is safely secured on jack stands before cranking on those 19mm bolts!

8. Remove the side panels near the engine lid to gain access to the top strut mounts (two Phillips screws and a 10mm bolt for the ground wire). If there's a rubber plug covering the top of the strut, remove it. A tip that I learned the hard way: Take a 19mm deep socket and loosen the nut on the top of the strut rod. Just loosen it enough to make removal easier later on.

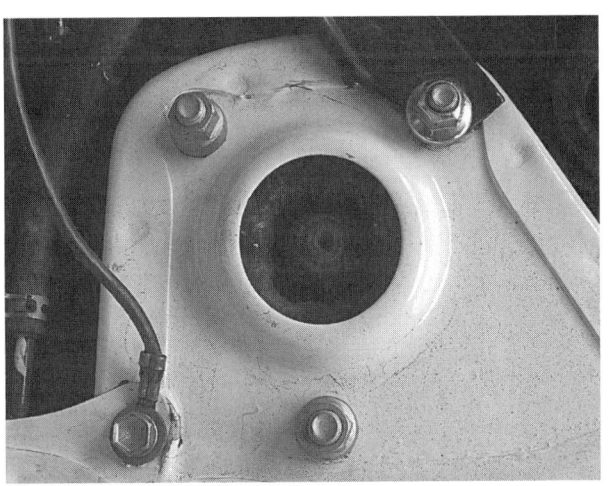

These 14mm bolts should not be nearly as difficult to remove, being relatively clean and protected from the elements. Take care here not to scratch the paint near the shock tower or your front fender.

9. Compress the strut to relieve pressure on the mounting bolts.

10. Once the strut has been compressed, you should be able to remove the mounting bolts. I used a drift to tap the remaining one out. I was then able to slowly pull the hub carrier towards me, working the bracket free from the strut.

Note: The strut can be removed easiest if you push the bottom of it towards the caliper side of the hub. It's fairly heavy, so be prepared to prevent damage as it come loose from the top mount.

Once the bolts are removed, the hub carrier should pull away easily from the strut body. A towel or other soft material can help protect these very heavy metal components from gouging one another.

You can place a couple of sockets and washers between the ears of the mounting brackets to prevent them from being bent when you tightened them in the vise.

11. Once the spring is compressed enough to be rotated by hand on its perch, you can remove the nut on the top of the strut rod, the 19mm one you loosened earlier.

12. Once the nut's been removed, there's a steel spacer below it. Slip the spacer out and put it aside. The Tokicos should include a new nut, but not the spacer.

A spring compressor allows you to squeeze the spring and safely remove the upper mount. The compressor will be more or less necessary, depending on the height of the springs being removed.

Now it's time to remove the 19mm nut on the top of the strut rod. It is also a good time to assess the condition of your strut tops.

13. If you are unlucky, the upper suspension mount might have welded itself to the strut rod, due to rust, corrosion or just bad luck. If the upper mount does not pop off, then you'll need to use brute force, or some other method of persuasion, whatever it takes. Trust me, it's simply there by the force of friction. Once the upper mount was removed, clean it off; you could even spray some black paint on it to make it presentable. Be sure you let it dry completely before reinstalling.

14. The next challenge is the large 2" nut that retains the internal strut mechanism. Assuming you

Ready for reassembly! Well, maybe after some serious cleaning. Cleaning all the nuts and bolts with a wire-wheel is a good idea. Now's the time.

do not have the correct tool to remove this (and trust me, it's on there TIGHT), here's a way to remove it: Lay the strut body in the vise horizontally to hold it more securely. Get a large pipe wrench on the nut and slowly crank it off. Be careful not to gouge the strut body too badly. I had to fit a long steel tube as a handle extender on my pipe wrench to apply enough leverage.

Wrap some protective packing material around the shaft to protect it in case the wrench slips during the reinstallation of the gland nut.

15. Once the nut is loose, reposition the strut upright in the vise, as it's filled with oil that will leak out when the nut is removed.

You'll need to add some lightweight oil between the strut cartridge and the strut housing to improve heat transfer. It doesn't take much, and Tokico recommends filling to within 1-1/2" to 2" of the top of the housing tube.

16. Tokico provides a new gland nut (I guess they know how badly a pipe can chew up the old one). They also recommend both a torque setting (~90 ft-lbs.) and a minimum spacing that describes the amount of threads showing. If this spacing is below the minimum, Tokico includes a spacer washer that should be inserted atop the cartridge before the gland nut is installed. It was not necessary in my case.

17. Install the suspension spring. Once the spring was in place, the upper suspension mount can be installed. Luckily, the nut on the strut rod (don't forget the collar!) requires a much lower amount of torque.

18. Reinstallation was simple with the exception of compressing the strut to install it into the hub carrier. During the removal, the strut is in the proper position, and compressing it is merely to take it apart. During assembly, you need to compress it, then position it, and without the proper tools, it can be frustrating.

With the rear complete, it's time to do the front. The fronts are mechanically similar to the rear, with some differences you'll see right away so grab your BGB and have at them.

Note: A special note on installing the three nuts on the studs at the top of the strut. The BGB calls for 59 lbs-ft. on these

It's time to install the spring. The top of the spring has a flat top and tighter coils.

Once the upper mount has been installed, your new strut/spring combination is ready to be reinstalled.

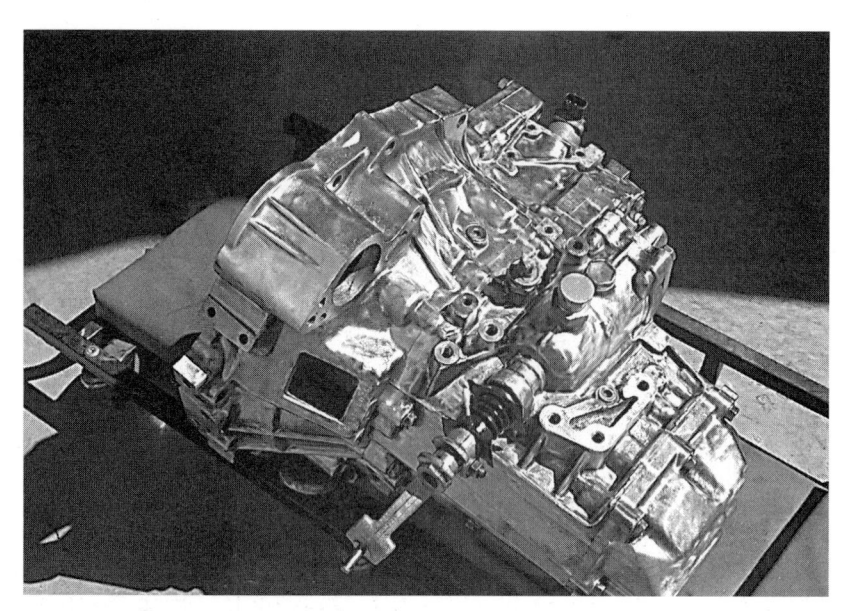

Toyota MR2 Performance

The E153 transmission is a worthy upgrade to any early (1991–1992) SW20. If you can't afford it, Redline MT-90 is your next best bet.

Mityvac is a brand name of hydraulic bleeding equipment that make brake-bleeding a one-man job. Speed Bleeders are an alternative. Photo courtesy Mityvac/Lincoln Industrial.

14mm nuts, which seems way too high. I tightened them to a little over 40 lbs-ft. Volume One of the BGB states a different torque rating: 47 lbs-ft. Given that the front struts call for only 36 lbs-ft., even the lower value seems more than adequate.

Upgraded Transmission Fluid

The 1991–1992 gearbox synchronizers are notchy, especially second gear, and this only worsens with age and mileage. Cable shifted transmissions like the MR2 will never yield a Honda or Mazda Miata feel, but early SW20s, especially those with some miles, and perhaps some impatient former owners, can be especially problematic in this regard. Improved transmission fluid can help.

This modification is even a good idea for 1993+ cars with the upgraded E153 transmission. Treat your transmission well, and it'll do the same to you. Redline MT-90 is the best choice here for your MR2.

Note: A quality aftermarket fluid will help the synchronizers do their job, and even possibly extend the life of your transmission. Be realistic with your expectations, though. If the synchronizers are significantly damaged, even the best fluids cannot repair metal.

Note: Redline MTL fluid is a 70W80 fluid. Redline also makes a thicker (more viscous) version of MTL called MT-90, which is a 75W90 fluid and will help with shifting action as well.

The 1991–1992 SW2x wheels were not only small but failed to compliment the visual lines of the car as well. The tall sidewalls only added to their visual delinquency.

Upgraded Brake Pads & Fluid

Keep in mind the range of use for your brake pads. Running track-only brake pads on the street can be a nightmare. Most aggressive pads are noisy, dusty, perform horribly when they're not track-hot, and chew rotors like crazy.

The good news here is that brake pads aren't hard to switch out. This not only makes swapping out track pads before you drive home feasible, but also—perhaps more importantly—enables fairly easy experimentation with various pads from various manufacturers until you find a pad that suits your needs. Porterfield R4S pads and ATE

SW2X STAGED BUILDUPS

The 1993–1995 wheel upgrade won't make a huge difference, but we're not looking for a huge difference here. These wheels offer a slight increase in handling stability, better range of performance, tire selection and improved aesthetics.

An upgraded sidemount intercooler will allow higher boost levels to be run than the stock intercooler, and offer more consistency from run to run as well. GReddy has been manufacturing the SW20 for years, while other small vendors like AutoLab have taken quality intercooler cores, like those from Spearco, and manufactured their own kits.

Type 200 or Super Blue is a nice setup.
Note: Use a less expensive DOT 4 fluid to flush the old stuff out before pouring in the $12-a-liter ATE Type 200.

Swap 1991–1992 SW2x Wheels for 1993–1995 OEM Wheels

The stock 195/60/14 and 205/60/14 size tires for the '91–'92 SW2x models have no place on a 2800-lb., rear-drive, mid-engine, turbocharged car that develops 200 lbs-ft. of torque. The fact that years later Toyota would bolt on a larger wheel and tire package on the lighter, far less powerful, normally aspirated ZZW30 is plenty of evidence that the smaller rubber needs to go.

While other adjustments were made at the time to presumably help stabilize the handling, the lengthening of the rear trailing arms for example, beyond toeing-in the alignment of the rear wheels, changing the tire size is the most accessible and inexpensive adjustment to increase handling stability and overall grip. The '93–'95 tire sizes (205/55/15 front / 225/50/15 rear) noticeably improve the handling, especially in the rear end.

Used OEM wheels from 1993–1995 MR2s (15x6 front/15x7 rear) are available for around $300 if you look around, and both normally aspirated and Turbo models share the same wheel and tire sizes, which should make a new set of wheels easier to track down.

Subjectively speaking, the later OEM wheels look quite a bit better than the early wheels as well. In lieu of OEM wheels, the aftermarket obviously will sell you just about anything you're interested in buying. Just keep in mind to retain the stagger in 1993–1995 width Toyota included front to rear, 225 rear, 205 front.

Note: The brake upgrade in 1993 will not fit beneath the early 1991–1992 wheels, so if you've got an early SW2x, you can fit later wheels, but not vice versa.

At the end of Stage I, we've done some careful revising of the MR2, but we're still running stock boost, on a stock IC, with stock fuel and stock engine management, on the stock CT-26 turbocharger. The boost gauge and 2.5"–3" exhaust, while not incredibly visceral at this stage, are key to future engine modifications, not coincidentally major themes in Stages II and III.

STAGE II
- Upgraded sidemount intercooler, shroud, SPAL fan, IC sprayer
- Aquamist 1S water injection
- KO Racing downpipe
- TRD strut tower bars
- TRD sway bars, 1993+ OEM sway bar end-links
- K&N air filter
- GReddy Profec-II
- ATS throttle body inlet
- ATS Racing CT27 Combo kit or:
- CT20B and HKS Fuel-cut Defense

Upgraded Side-Mount Intercooler, Intercooler Shroud, Fans, Intercooler Sprayer

Before we turn up the boost, you need to install a boost gauge if you didn't already as outlined in Stage I. In Stage II, one of the first things I recommend is an aftermarket sidemount intercooler like the Spearco-built unit from AutoLab, or the

129

Toyota MR2 Performance

The stock intercooler uses a fan as well, but it is relatively puny and does not turn on until engine bay temperatures reach 140°F.

Any intercooler is susceptible to heat soak; a relatively small air-to-air intercooler mounted in a tiny hole in the side of an engine bay is even more susceptible. Enter the intercooler shroud and Spal fan—and perhaps even an intercooler mister to help with cooling.

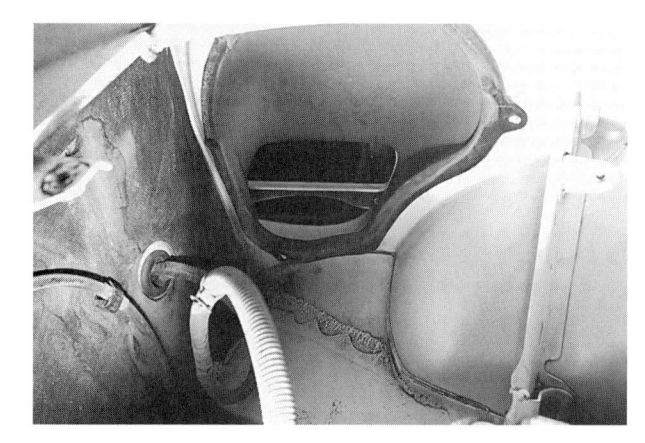

One of the few drawbacks of its mid-engine design: Rather than a huge front-mount intercooler being coated with a wall of fresh air, the MR2 intercooler gets fresh air through this restrictive aerodynamic duct as well as hot air from the turbocharger, which it must cool.

The HFS-1 water-injection kit by Aquamist is an excellent entry-level system that allows for injection response to be initiated by either a predetermined boost level or injector duty-cycle. The HFS-1 has a DDS3v8 fail-safe built in, automatically switching to lower, safer boost settings when less-than-ideal water flow is detected. Photo courtesy Aquamist.

comparable intercoolers from GReddy or Berk.

Side-mount intercooling is the best bet for most engines running under 300 whp, all things considered. In regard to cost, simplicity and weight, they work fairly well. These setups do, however, have limitations. We discussed intercooling in Chapter 4, regarding issues like heat soak, and ways to supplement the effectiveness of an aftermarket sidemount air-to-air intercooler. In a nutshell, to help the intercooler recover more quickly, and to increase its overall effectiveness, two fans (one pusher, one puller), a shroud and an IC sprayer, when combined with water-injection should provide effective intercooling for your MR2 for this stage.

Other options include larger trunk-mounted intercoolers, and more complicated air-to-water units as well. These units aren't absolutely necessary at this point, but if you are building a 350+ whp MR2, you might consider an amended schedule for your buildup.

Note: Keep in mind that intercoolers do not see much wear. An aftermarket sidemount can always be used for now, and sold later to help finance a trick setup if the need develops.

Water Injection

As discussed in Chapter 2, water injection really is a faultless modification beyond its considerable initial expense. Whether you're tuning with it for power, or, more likely at this stage simply adding it for durability, water injection is a strengthening modification that has a place on almost any forced-induction application. Its detonation suppression is

SW2X STAGED BUILDUPS

The trunk on the AW11 and SW2x makes for a very convenient place to store your water tank. Photo courtesy Vespremi/Crites/Smith.

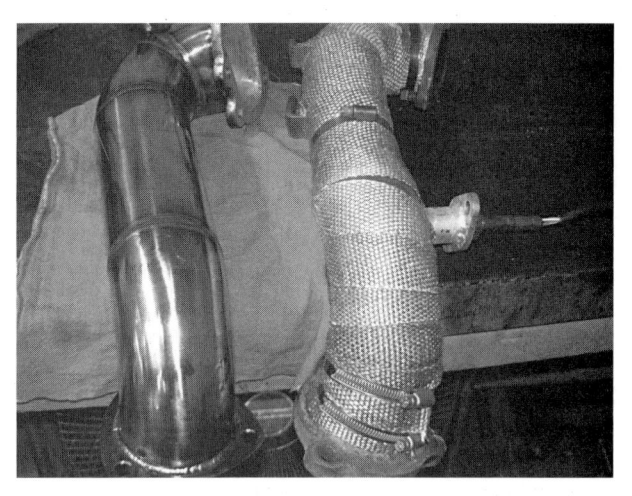

Being bolted directly to the business-side of the turbocharger, the downpipe sees an incredible amount of heat, as EGTs can rise well over 1000°F. Wrapping the downpipe is a way of keeping heat in the downpipe to encourage exhaust scavenging, and keeping it out of the engine bay. Such wrap is a budget alternative to the superior process of proprietary high-temperature coating from Finishline Coatings, Jet Hot, or Swain. Note the O_2 sensor bungs on the downpipes. Photo courtesy Jensen Lum.

Both Jeff Fazio and David Vespremi's (pictured) highly developed SW2x MR2s utilize the intercooling and detonation-suppression abilities of water-injection products. Photo courtesy Vespremi/Crites/Smith.

extremely valuable.

How exactly you use your water-injection system will help you make many decisions right off the bat. This is a complex issue worthy of an entire book on its own. What system should you get? What size nozzle should you use? Where should you mount the water tank? What mix of ethanol and water is best for you? Should you use distilled water, or is tap water acceptable? Should water injection be used on a fresh rebuild while breaking it in? Are there really quality differences between water-injection kits? Can't you just make your own water-injection kit? Your best bet to answer specific questions like these and more is to join a water-injection-specific forum, or a 3S-GTE/MR2 forum with a sub-forum that addresses water-injection information sharing. Additionally, communities like the Subaru WRX and various turbocharged Mitsubishi models have active water-injection users who are able to provide data, anecdotes, and other information to help you in your research.

Aquamist makes a very high quality kit, and assuming you're not running a standalone EMS, the Aquamist HFS-1 is your best choice here. The wonderful thing about Aquamist water-injection systems is that they are upgradeable, allowing you to upgrade your system later should the need arise.

Downpipe

There is a bell-mouthed catalytic converter—or "cat"—right off the turbocharger. It is a heavy thing filled with a smothering honeycomb material, barely letting visible light through, much less 300 horsepower. Replacing it with a 2.5"–3" coated downpipe is what we're looking to do here.

Air Filter

Traditionally, intake air filters are among the first changes many enthusiasts reach for. At stock boost levels, though, a new intake filter is not a necessity. Though they may provide around 5–8 whp, most intake systems will sacrifice the filtering efficiency the stock airbox provides, and can be guilty of increasing air intake temperatures by removing the barrier the stock airbox provides. Leaving the stock airbox intact until you're ready to really start increasing the boost is a perfectly acceptable, and

Toyota MR2 Performance

What separates application-specific air intakes from universal applications is mainly the length of the tube, location and number of vacuum line dimples, adapter style, and angle of the coupler bends. This is actually the MAP version of the KO intake, for Gen III 3S-GTEs, and Gen IIs converted to speed density/MAP. MAP means no AFM, and no AFM means no AFM adapter is necessary. Photo courtesy KO Racing.

The HKS FCD is a relative dinosaur in the MR2 aftermarket, having been used since the 1990s to deal with the factory fuel-cut. When the fuel-cut engages, it is adjustable by the turn of the dial in the center. Photo courtesy HKS USA.

Many tuners go with the largest air filter that will fit, and there is little to fault with that approach. This KO Racing intake locates the air filter as far as possible toward the driver's side intake vent. Photo courtesy KO Racing.

appropriately frugal approach, and one I recommend.

Many shops like KO Racing offer all of the necessary parts to fabricate your own intake. An air intake system simply consists of a pipe to connect to the turbocharger, vacuum dimples, an adapter to connect to the AFM and a filter to go on the end of the AFM. That's it. Your primary concerns are keeping the air filter free from water when it rains, and heat when you're driving and idling. For this reason, the air intake box by ARC is also highly recommended, but as with many trick parts, it's expensive, hard to find, and fitting it to the Gen II (it's manufactured specifically for the MAP Gen III 3S-GTE) requires some custom work. Overall, the best choice here is a custom KO Racing (or similar) intake pipe coupled with a K&N filter.

Fuel-Cut Defense

An HKS FCD is the simplest way to deal with the factory fuel-cut right now. The difference between the GReddy BCC and the HKS FCD is that the GReddy BCC completely removes the fuel-cut, while the FCD from HKS simply allows you to delay the fuel-cut engagement.

Boost Control

We're almost ready to turn up the boost. Whether you go with a simple, cheap manual ball-and-spring boost control or an electronic boost control depends on your budget and application. If you can afford it, electronic is the way to go.

It is true that some electronic boost controllers are overly complicated— units from APEXi come to mind, offering what, on paper, looks very promising: gear-dependent boost control programmable against rpm. While well-intentioned, the APEXi AVC-R is an electronic boost controller that tends to get in its own way. It is difficult to set up, the install requires splicing into the OEM wiring harness if you intend on using all of its given functions, and it is prohibitively expensive, with an MSRP of over $600.

If you want the cost-effectiveness and simplicity of a

SW2X STAGED BUILDUPS

The GReddy BCC completely removes the fuel-cut, which sounds convenient but could prove costly if you ever lose a critical vacuum line and boost levels skyrocket. Photo courtesy Brian Hill.

The best strut tower bars not only connect the strut towers together, but brace again to a firewall as well, like this bar from TRD. Note the manual boost controller zip-tied to the bar and the Zen blow-off valve beneath it.

manual boost controller, a Twosrus MBC is a strong choice. Nice, simple electronic units like the GReddy Profec B II or Blitz Dual SBC are solid choices if you want to be able to adjust boost levels from the cockpit, while other units from other manufacturers offer additional (and simple) options—built-in boost gauges and turbo timers that can add tremendous value. The cost of smaller modifications like these adds up, so save where you can!

Strut Tower Bars

Chassis bracing is always a good thing, especially on sunroof and T-top versions that make up the majority of MR2 production. One of the my personal favorite design details on the SW2x, beyond the neat little courtesy lights on the interiors of the doors, is the rear strut tower bars. I'm a huge fan of suspension and chassis braces: strut tower bars, lower-tie bars, roll bars, bars for braces and braces for bars, you name it, I'm on it. However, something that never made much sense to me is tying one apparently unsturdy shock tower to another. It stands to reason that the best strut tower bars would connect to a firewall or some other less-shaky piece of equipment—and that's where the MR2 strut-tower bars come in.

The SW20 has a factory X-brace at the rear, providing a solid backbone of some geometric significance. The best alternative for aftermarket bars, made by TRD, go one step further, not only "X-ing" across the engine bay and bolting to the firewall, but adding a third bar that connects the strut towers as well. TRD strut tower bars are also available at many Toyota dealers and aftermarket vendors, so they are easy to find.

Adjustable aftermarket sway bar end-links are recommended once you start running thicker aftermarket sway bars. They are sturdier than the factory links, and allow for proper preloading of the sway bars during installation.

Adjustable Sway Bars, Adjustable Sway Bar End Links

In addition to the traditional benefits of stiffer, adjustable sway bars, it is also a common trick for SW2x stock-class autocrossers looking to increase traction to the inside rear wheel to add a massive sway bar in the front—acting very much like a poor man's limited slip differential. A set of TRD sway bars front and rear, ideally with aftermarket sway bar end-links, is an excellent choice. ST sway bars are also strong options as well.

Note: Keep in mind that under extreme use, massive front sway bars like the bar from ST can rip

Toyota MR2 Performance

[ATS Racing dyno chart]

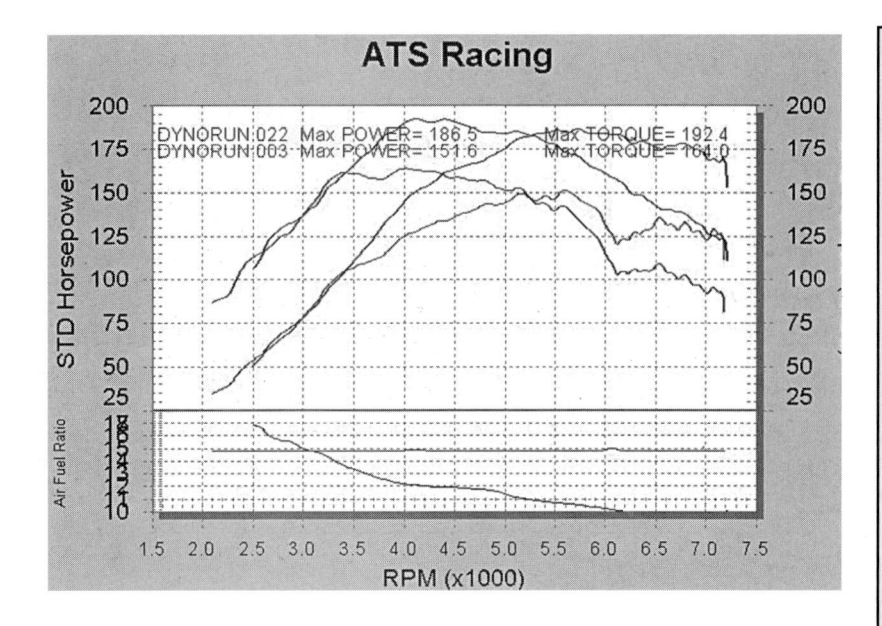

This dyno comparison shows a stock CT26 dyno run against a CT27 at stock boost levels. The result? An increase of 35 whp and 28 lbs-ft. Notice how early in the powerband the CT26 begins to fall off, compared to the CT27 powerband staying flat much longer.

The CT20B is available with both steel and ceramic wheels, and maintains the dual-entry function of the CT26. Like the CT27, the CT20B will extend the powerband into the rev limiter without other system changes larger turbochargers might demand. Photo courtesy Jensen Lum.

the sway bar mounts right off the car, so if extreme use is what your MR2 is going to see, preemptively reinforcing the front sway bar mounts is necessary.

Note: As coil-overs can change the length of sway bar end link you need, if you are indeed running a coil-over suspension, verify the length of end link you require.

CT27 ALTERNATIVE: CT20B

While not available in any junkyard in the United States, the CT20B represents a simple fix for the powerband because there is no (immediate) need for larger injectors, ECU remapping, or even replacing the catalytic converter. It has an extra stud on the downpipe/cat flange, but you can simply remove that stud, and the stock cat or your aftermarket downpipe will bolt right up. Its ultimate performance may just fall short of the CT27, but for most owners, the performance will be extremely similar.

Finding clean, mechanically sound CT20Bs isn't always easy, so don't overpay for one: the

Among the many revisions Toyota thought necessary for the Gen III version of their 3S-GTE was a more potent turbocharger, but one that still allowed for a tractable, seamless powerband. The CT20B is what they came up with. Photo courtesy Jensen Lum.

price should always be less than $1000, closer to $750.

Keep in mind that planning ahead can save you some bucks down the road. If you're sure 275 whp is plenty, a CT20B, or the recommended ATS Racing CT27, are ideal choices. If you think though that big power is in your immediate future, don't even bother with the turbocharger at this point. Other things need to be replaced first.

ATS Throttle Body Inlet (1991–1992 SW20 only)

As with most other airflow modifications, results of this modification can be anticipated in terms of percentages: What might make only 5 horsepower at 200 might make closer to 10–12 at 400. Though the relative horsepower offering is slim, so too is the cost: Under $100 for around 5 whp and sexy billet aluminum isn't a bad deal.

SW2X Staged Buildups

A Stage II SW20 isn't as extreme or complex as a Stage III SW20, but is far more focused and rapid than a simply upgraded Stage I SW20. Establishing your development plan for your MR2 as early as possible is important, especially so as the modifications increase in number, cost and complexity. Photo courtesy Brian Hall.

This supercharged MR2 could be bought—and tracked several times—for the cost of reliably moving from 275 whp to 400 whp. Perspective is important in any buildup. Photo courtesy Bryan Heitkotter.

CT27 Combo Package

Turning the boost up on the CT26 turbocharger will increase the torque at lower rpm where the turbocharger can still flow enough air to do its job, but the car still won't pull to redline like any proper sports car should. An easy solution here is the aforementioned CT27 combo kit from ATS Racing.

The CT27 combo kit from ATS Racing comes with the CT27 turbocharger, tuned ECU, fuel pressure regulator and 170°F thermostat, with a downpipe, intake, boost controller, and 3" exhaust necessary for the kit to do its job. With this kit, 17 psi should have you approaching 275 whp.

Note: A Stage II version of the CT27 kit is available, which adds a modified fuel rail, Walbro 255hp fuel pump, and an increased potential of 20–40 whp.

The 275 Horsepower Stopping Point

Stage II is the most comprehensive stage of development of an SW20. It elevates your MR2 beyond the simplest bolt-ons, and adds a degree of adjustability and durability that isn't available in Stage I—and in some ways, Stage III either.

A Stage II SW20 is very rapid, possessing impressive grip and very responsive handling. It has also addressed all of the stock areas needing improvement. The Stage II MR2 Turbo is capable of embarrassing cars costing many times its price, offering 12-second quarter-mile potential, approaching 1g of lateral acceleration and brakes that can cause nosebleeds. It also can be remarkably civil and drivable around a city. In many ways, a Stage II MR2 is the ultimate MR2.

At this point we're also coming up fast on 300 whp. At these power levels, everything changes, and more horsepower becomes more expensive and complicated. The stock 440cc injectors are already maxed, and must be replaced; in fact, even 550cc injectors are running on borrowed time. The OEM ECU alone is not capable of controlling even larger injectors, so some method of controlling the bigger injectors must be found. If you haven't already replaced your downpipe, at this point it is a borderline necessity—same with the intercooler and intake. The stock fuel pump is quickly becoming taxed as well, and will also need to be upgraded with a Walbro unit. And the clutch, while sturdy up until now, is best upgraded as well.

If you plan on launching the car aggressively, such as for drag racing, there are also other concerns, mainly, axles/joints. Being a mid-engine, rear-drive, turbocharged car, the MR2 is capable of launching very hard if set up properly. Even on street tires and a stock turbo, low 1.7–1.8 second 60-foot times are possible, and that sort of "hook"—the ability for the rear tires to grab and launch the car instead of spin—puts tremendous strain on the clutch, axles, and inner CV joints. Somewhere between 300 and 350 horsepower, these pieces become vulnerable.

Cresting above 300 whp is a whole new ballgame whose pursuit can easily exceed the price of a second MR2, a nice red AW16 with T-tops, for example. Because of this, somewhere around this output level is a natural stopping point for a lot of SW20s out there. This is not a bad thing. The SW20 is a well-balanced and capable machine that becomes a downright handful beyond 300 whp.

135

Toyota MR2 Performance

The APEXi Power FC is plug-and-play for the Gen III 3S-GTE, but requires a custom harness for use on the Gen II, available from aftermarket vendors. It still represents a solid choice in powerful engine management, highlighted by use of a handheld commander to input adjustments. Photo courtesy Vespremi/Crites/Smith.

The TEC-series from Electromotive offer powerful ignition components for 3S-GTE owners, unique from other engine management system.

Of course one of the bonuses of a turbocharged car is that boost can always be turned down for less horsepower, but why build a car beyond its sweet spot? Before you spend thousands of dollars building a 400 hp SW20, you should find one and drive it first. Big turbos hitting mid-sweeper in an MR2 might be an experience best left to you and your PlayStation 3.

STAGE III
- ACT clutch
- Hydra Nemesis engine management
- Larger injectors, fuel rail and fuel pump
- Upgraded turbocharger
- HKS 264 cams, adjustable cam gears
- Kaaz 1.5-way LSD
- 215 front/235 rear tires on 17x7 and 17x8 wheels
- ATS Racing CV joints
- Extrude-honed factory intake manifold, phenolic TVIS plate
- KW variant 3 coil-overs

At this point, a standalone engine management system is an absolute necessity, the tuning of which comes via a wideband O_2 sensor. A wideband O_2 sensor operates like a standard sensor, but with incredibly high resolution and reaction, able to report back immediate information on air/fuel mixtures, allowing for precise tuning.

Clutch

Like many factory components, the factory clutch is surprisingly capable until around 300 whp. Not any more. Unless you're going after huge horsepower (or more precisely huge torque), an upgraded clutch disc and pressure plate should perform fine, with no unsprung, or "pucked," clutch-work necessary.

Normally overbuilding is the theme, but the more holding power a clutch has, typically speaking the less driveability it possesses, so the more precise you can be here with your power goals, the better off the driveability of your MR2 will be. Don't buy a clutch for the horsepower you're hoping to get at some point in the future; buy your clutch with the subsequent twelve-month window in mind. An ACT rated for your specific power/torque level is a strong choice here.

THE HYDRA NEMESIS EMS

The Hydra Nemesis is a well-supported, powerful EMS for the 3S-GTE. Rivaling systems from multimillion dollar manufacturers like AEM, the Hydra Nemesis is strongly capable of managing the combustion process in your MR2 Turbo.

One of the most notable features, beyond the sheer power of the hardware and software, is the availability of a plug-'n-play harness. Engine management systems basically requiring rewiring the way your engine runs, not an easily digested thing for many owners. A harness that allows the unit to basically plug right in (and software that allows it to actually start), go a long way to making this system attractive for those in need of its power. Bells and whistles like launch control and full control of accessories like power-steering and ABS should not be overlooked or underappreciated either.

The Hydra Nemesis 2 product line is powered by dual 16-bit 25MHz Motorola processors which provide fast and accurate control of your engine. The base Hydra Nemesis 2.6 plug-and-play engine management system comes with everything you need to install and run the Nemesis stand alone engine management system on an MR2 Turbo. The system has the following features:

This exhaust manifold is probed with four exhaust gas temperature probes. EGT probes give critical data to engine management systems. One in each cylinder allows a closer monitoring of what is happening in each individual cylinder, rather than the less-precise method of a single probe taking an average reading—in the downpipe, for example.

- Easy installation on an SW20
- Eliminates the stock airflow meter (AFM) and cold start injector
- User configurable 32x32 fuel and timing maps with selectable rpm and load points
- Two selectable 32x32 fuel and ignition offset maps for alternate fuels
- Directly supports the NGK L1H1 and L2H2 5-wire wideband oxygen sensor (additional cable required)
- Uses stock engine sensors including the distributor
- Launch control (two-step) and flat shift for drag racing applications (additional cable required)
- Controls the stock ignition or MSD 6 ignition setups
- Runs full sequential injection
- Supports full sequential ignitions (additional igniter adapter required)
- Operates all accessories including AC, power steering, cruise control and ABS
- Controls the TVIS, EGR, fuel pump volume and engine check light
- Uses the stock idle speed valve to provide a stable idle under different loads
- Internal 4-bar MAP sensor supports up to 45 psi boost
- Rev limit to 16,000 rpm with adjustable soft (random miss) and hard (complete cut) rev limiters
- rpm adjustable knock detection threshold for detonation protection
- Adjustable fuel cut for overboost protection
- Jumper configurable to work with 91-92, 93-95 and Gen III 3S-GTEs
- Uses stock narrowband oxygen sensor to maintain proper fuel mixture while cruising and idling
- Individual cylinder fuel and timing advance trims
- Quick connect two-piece laptop cable can be easily routed to passenger cabin
- Windows-based tuning software (requires a laptop with a serial interface)
- On-board and laptop data-logging at adjustable timing rates
- Controls Aquamist 1S and 2C water injection setups
- Adjustable air and coolant temperature correction of fuel and timing
- Supports staged injection setups (additional injectors and cable required)
- Controls boost (boost control option required)
- Operates as a turbo timer (turbo timer option required)
- Comes with base maps to start most engine setups and drive them to the dyno for proper power tuning
- Controls any level of power from a bone-stock 170 whp up to +600 whp monsters

Replacing fuel system components is not a process to be taken lightly. Safety must be the highest priority in any system designed to move large amounts of gasoline in and around a hot engine, and the highest quality components should be used. This is not the area to save money. The very nice WolfKatz 3S-GTE fuel rail is shown.

Somewhere in the neighborhood of 260 whp for the Gen II 440cc injectors, and 300 whp for the Gen III 540cc injectors, the injectors are approaching their limits, unable to provide enough fuel even as they approach 100% duty cycle—basically wide open. Time to upgrade. Photo courtesy Jensen Lum.

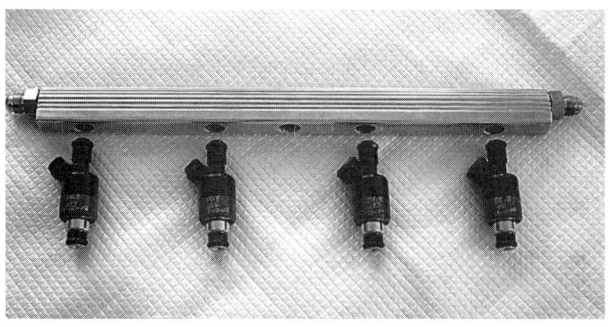

The factory Gen II 3S-GTE side-feed fuel rail is sufficient if you're still running a stock fuel pump and stock injectors. As you begin to upgrade these pieces, the stock fuel rail, with its tiny bore, should also be replaced. A Garage Advance fuel rail and 65 lb/hr injectors are shown.

Aftermarket Engine Management Systems

We talked about the basics of engine management systems in Chapter 3—the point here is that the stock engine management system, the OEM ECU, is no longer capable of optimally orchestrating the amount of horsepower Stage III is capable of. Most of these units have a niche—the Electromotive's use of advanced ignition components, for example—but most systems do the same thing: they run your engine with extreme precision, while offering a limitless amount of adjustability to suit your boost, your cam profiles, your powerband, your local octane supply, etc.

The choices are many: Hydra Nemesis, AEM, APEXi Power FC, Autronic, Haltech, Motronic, TEC-series by Electromotive and others are all available here. A Hydra Nemesis is a choice difficult to fault (see sidebar).

Note: Like many engine management systems, the Hydra Nemesis software features selectable fuel and ignition maps, one standard premium fuel map, and one for race fuel for example, and allows for fuel tuning by individual cylinder.

Fuel System Upgrade

Bigger Injectors—As we discussed in Chapter 3, a common upgrade for those running higher boost are 540cc/550cc injectors. This is around 25% larger than the Gen II 3S-GTE stock 440cc units. Rough math would indicate that they would then be suitable for another 25% horsepower—so then, if the 440ccs maxed out around 260 whp, the 550ccs from the Supra would be good for another 65 whp, getting you to around 325 whp at approaching (a nonideal) 100% duty cycle.

This rough estimate isn't too far off—for now, we'll say it is safe to assume that the 550cc injectors will get us at least to our 275 whp stopping point, and probably safely to 300 whp as well, which is right about where a Stage III SW20 starts wrapping up engine development.

While 300 whp does not sound as extreme as the

SW2X STAGED BUILDUPS

BASIC FUEL RAIL AND INJECTOR RECOMMENDATIONS
- Stock up to 260 whp: 440cc, stock side-feed rail
- 260–310 whp: 550cc, WolfKatz side-feed rail
- 310–400 whp: 800cc, SARD side-feed with WolfKatz side-feed rail or 800–900cc top-feed injector, WolfKatz top-feed rail
- 400+ whp: WolfKatz High HP MR2 fuel system

BASIC FUEL PUMP RECOMMENDATIONS
- Stock up to 275 whp: stock fuel pump
- 275–400 whp: Walbro 255 lph or OEM 2ZJ-GTE (JZA80) Supra fuel pump
- 400+ whp: WolfKatz High HP MR2 fuel system

Note: These are basic recommendations. Always monitor air/fuel ratios, fuel pressure, injector pulse width and other parameters to be certain you have adequate fuel supply.

Stage III label might suggest, as we've discussed, 400 whp really pushes the boundaries of the SW2x platform. In the hands of a skilled driver, incredible performance can be extracted by a 400 whp MR2, but in regards to general ownership, there are significant issues. In truth, most owners end up developing a Stage IV engine, and the rest of the car is stuck back somewhere in Stage I or Stage II.

Note: Many injectors are rated by lbs./hr, rather than the metric cc value. To determine lbs./hr when given ccs, divide the cc value by 10.5. In this case, a 440cc injector would be rated at 41.9 lbs./hr, an "880cc" injector then around 82 lbs./hr, etc.

Fuel Pump—In Chapter 3 we discussed the basic concepts behind adequately fueling a modified SW20. These issues—including your personal tuning philosophy (regarding fuel pressure, injector duty cycle, and basically safe, methodical tuning vs. pushing the envelope, and the tuning philosophy of the person performing your fuel and timing mapping) all make assigning a horsepower rating to a fuel pump sketchy at best. Ultimately, the only way to know for sure if you have adequate fuel supply is to consistently measure air/fuel ratios and fuel pressure at various boost levels, including the maximum reasonable boost level you expect to run, and periodically checked again with any changes in modifications that might affect airflow. See the sidebar above for more recommendations.

Turbocharger Upgrade

This is perhaps the most critical step you will take with your SW20 in regards to giving it a personality. A turbocharger can completely change the way you think not only about your 3S-GTE, but about automotive engine design overall.

I owned an SW20 for three years in the 1990s before I ever even considered replacing the turbocharger. A $2,000 turbo kit seemed like too much to me. Others needed that much power, but not me. I spent most of my time on suspension setup and watching—with considerable

Adjustable fuel pressure regulators allow you to—you guessed it—adjust the fuel pressure in the rail that the injectors are ultimately fed by. This can give your injectors some elasticity in their application, allowing higher fuel pressures for tuning support.

The 2ZJ-GTE (JZA80) Toyota Supra is a common source of parts for SW20 owners—well, injectors and fuel pumps anyway. This Supra fuel pump, or the Walbro 255 lph, will supply the fuel needs of most modified SW20 owners in their quest for horsepower.

Toyota MR2 Performance

MAKING HORSEPOWER: A CONCERT OF WELL-CHOSEN PARTS

If making horsepower was easy, everybody would do it. The following setups are samples to illustrate what could be considered appropriate setups. The first column could be considered Stage I or II, the middle Stage II or III, while the latter is firmly Stage III.

CT26	CT27 or CT20B	T3/T4 50-trim or TD06
230 whp	280 whp	320 whp
440cc side-feed inj	550cc side-feed injector	800cc top-feed injector
Stock fuel rail	WolfKatz side-feed rail	WolfKatz top-feed rail
Stock fuel pump	Supra or Walbro 255 lph	Supra or Walbro 255 lph
Stock FPR	Stock FPR	Aeromotive 13109 FPR
Stock ECU	ATS ROM-tune	Standalone EMS

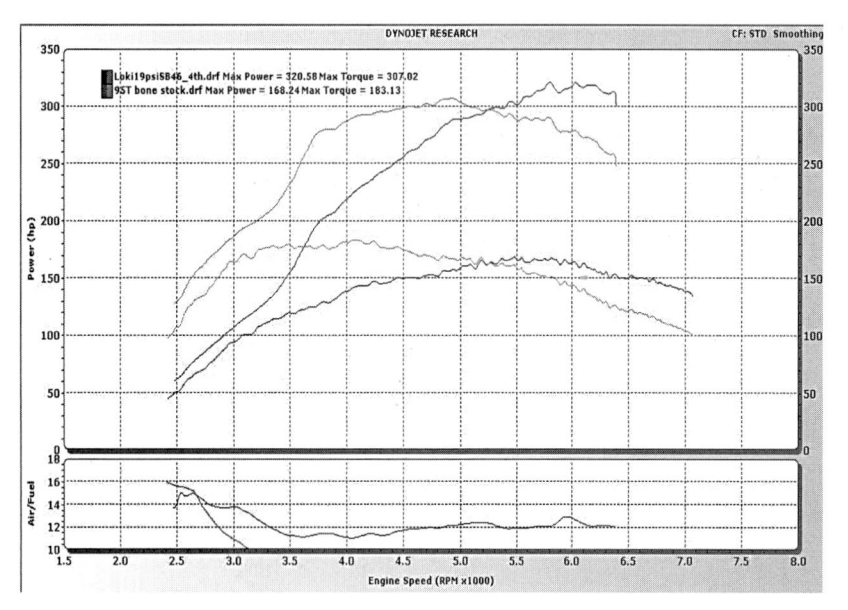

A dyno chart of a KO Racing T3/T4, 46-trim "Street Brawler" kit at 19 psi, graphed against a stock SW20. Stock: 153 whp, 183 lbs-ft.; T3/T4: 321 whp, 307 lbs-ft., almost doubling the factory horsepower, and improving the powerband everywhere. Image courtesy KO Racing.

amusement—others rip out their turbochargers for the next big thing from Turbonetics or HKS, always breaking something, and spending huge amounts of money, always seemingly unsatisfied no matter how much horsepower they made.

I never really appreciated the way a turbocharger can change a car's nature, though, until I drove a GReddy TD06 (20G)-equipped SW20. With the more efficient compressor of the TD06, no longer did the MR2 seem to limp towards redline; instead it screamed and wailed hell-bent for the rev limiter in every gear, and any boost level, every time. The engine literally felt twice as big as it had stock. Why? As discussed in Chapter 3, the answer lies in turbocharger efficiency. Matching an efficient turbocharger to the airflow needs of your 2.0 3S-GTE will yield a powerful, satisfying powerband.

The bottom line is that you need to consider what you want your MR2 to feel like. Where should the boost come on, and how hard do you want it to pull once it does? Do you want a light-switch—a car that has literally no power until the turbo comes on boost and hits its stride? Do you like the tractability of a quick-spooling turbo very much like the CT26 but with stronger upper rpm performance, or is downshifting for power part of the fun? Do whatever you can to physically get behind the wheel of other MR2s with various turbocharger upgrades to help you figure out what you want. This is a big, expensive step that cannot be properly measured simply by eyeballing dyno-charts or peak horsepower figures.

Learning to read a compressor map—if one is available for the turbocharger you are considering—is the best way to determine appropriate turbocharger sizing for your MR2, in conjunction with dyno graphs of similarly equipped MR2s, and talking to knowledgeable speed shops. The lazy man's way is to find a message board and ask somebody else what turbo kit you should buy.

The following pages include basic product information about a few of the best all-around performing turbocharger kits available for the 3S-GTE. The horsepower range the turbocharger will operate most naturally at depends on a wide variety of factors, namely boost level, airflow in and out of the head via manifold, camshaft and valvetrain upgrades, and the all-critical engine management tuning. To gain a rough understanding of where these turbochargers come in the pecking order, I have assigned each a sweet spot so that you may begin to establish the order of things.

SW2X STAGED BUILDUPS

The CT20B is Toyota's own answer on how to upgrade the CT26, earning a spot on the revised, Gen III 3S-GTE. The most dynamic difference between the performance of the turbochargers lies in upper rpm airflow potential: The CT20B will continue pulling all the way into the rev limiter. Photo courtesy Brian Hill.

The significant external difference between the CT20B and CT26 is the added stud on the turbine housing, the CT26 having 6, the CT20B7. Remove this stud, and the CT20 will bolt to any existing CT26 cat or downpipe, which can save you some money. Note the continued twin-entry design. Photo courtesy Brian Hill.

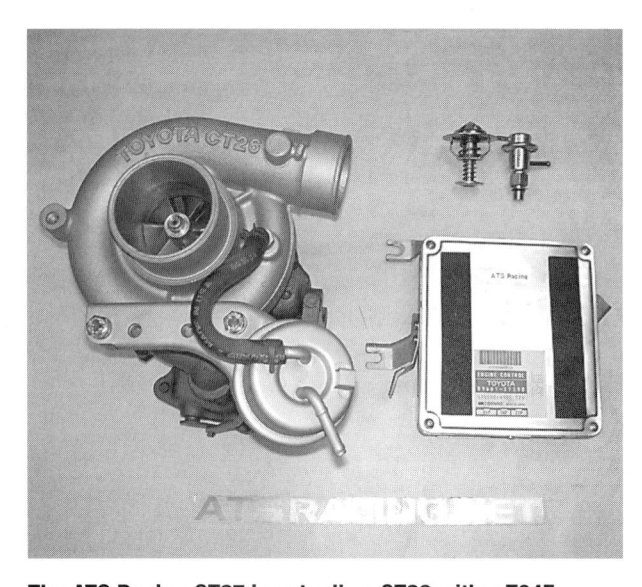

The ATS Racing CT27 is actually a CT26 with a T04E compressor wheel and increased exhaust housing flow. It is also the first CT26 upgrade to actually deliver serious performance, capable of over 300 whp with supporting modifications. The combo kit from ATS Racing is shown, which includes an ECU upgrade, ATS fuel pressure regulator and 170F thermostat. Photo courtesy ATS Racing.

Stock Toyota CT26 (Gen II 3S-GTE)—Performance Sweet Spot: none.

To be fair, the CT26 is a decent turbocharger for some owners. It's a great commuter turbo—spools quickly, allowing you to squirt in and out of traffic. It can make respectable torque, and is also free. The JDM ceramic CT26 is a bit better, and spools very, very quickly, making a strong autocross turbocharger, though even in that case, a CT20B would arguably be better. The CT26 falls on its face by 5500 rpm, which keeps it out of any sweet spot, especially a wonderful 7250 rpm redlining, 2.0L champion of an engine like the 3S-GTE. If you want anything more than "quick commuter" out of your MR2, ditch the CT26; you've got many affordable options. And don't waste your money sending it out to have wheels changed out, or turbine housings ported. Start over with something else entirely.

Stock Toyota CT20B (Gen III 3S-GTE)—Performance Sweet Spot: Stock–285 whp.

If you haven't figured it out already, the CT20B is an inexpensive, potent stock turbocharger cribbed from the Gen III 3S-GTE. Being an OEM turbocharger, it is oil and watercooled, just like the CT26. This turbocharger is highly recommended in almost any modest, under-300 whp application: street, mountain roads, road racing, autocross, or drag racing. Very sweet street turbocharger.

ATS Racing CT27 Combo Kit—Performance Sweet Spot: stock 285 whp. An excellent alternative to the CT20B, ATS Racing's CT27 kit seems less detailed on paper than other kits, but that's because it doesn't bother to reinvent the wheel. The CT27 is basically a reworked factory (CT26) turbocharger. If you don't require big horsepower—and you're sure about that—the CT27 is difficult to fault. An advantage the CT27 has over the CT20B is the security in knowing that you are buying a fully rebuilt turbocharger from an established retailer. Finding a clean, used CT20B is feasible,

141

Toyota MR2 Performance

Improving upon the proven 20G-design foundation of the GReddy TD06SH kit, ATS Racing created their own version of the kit, including a choice of TD05 or TD06 turbocharger, a downpipe that is braced to help prevent cracking, a 38mm external wastegate to prevent boost creep, and every single line or fitting necessary to install. Photo courtesy ATS Racing.

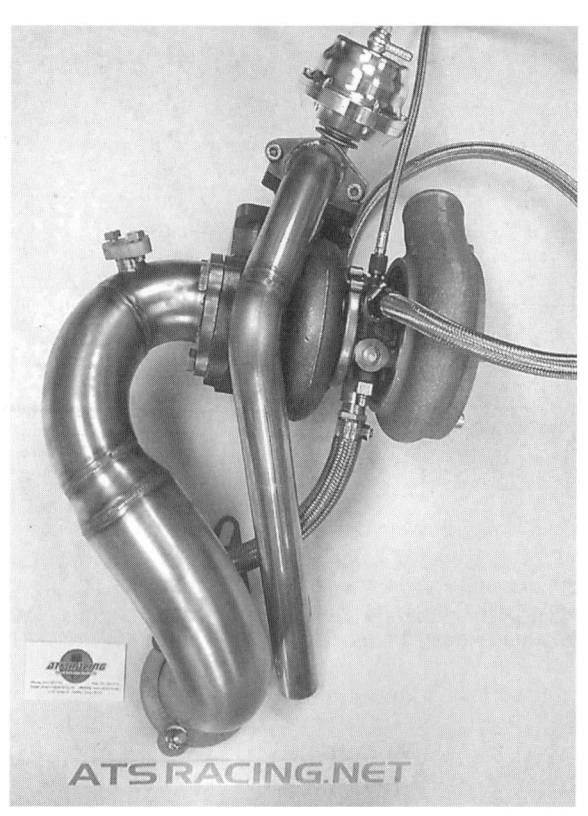

Another important detail in the detail of the ATS Racing TD06 kit was the inclusion of "water-cooling" capability, allowing coolant to supplement the oil-cooling-only capacity of the original GReddy TD06SH kit. Note both the stainless steel oil and coolant lines, and the external wastegate dump tube that can remain vented, or, better yet, routed back into the exhaust system pre-muffler. Photo courtesy ATS Racing.

but finding a clean CT27 is as easy as calling ATS Racing. It is an excellent all-around turbocharger, even able to go beyond 300 whp with related enhancements, though keep in the mind of concept of "sweet spots." If you want 375 whp, keep looking.

Among other proprietary pieces, the combo kit adds an ECU upgrade to the CT27 turbocharger itself that can be reprogrammed to control larger 550cc injectors to take full advantage of the CT27's horsepower potential.

Note: 3" exhaust, downpipe, aftermarket intake, and boost controller are highly recommended. Dual-map option to run race gas for increased horsepower.

ATS Racing CT27 Features and Specs:
- ATS CT27 (custom specific)
- Tuned Rom (ECU)
- ATS Fuel Pressure Regulator
- 170°F thermostat

ATS Racing TD05 or TD06 Kit—Performance Sweet Spot: 275–325 whp. In addition to the CT27, ATS Racing also offers TD05 and TD06 kits, making ATS and KO Racing two of the most SW20-dedicated companies in the United States.

A version of the TD05 was most notably first offered by HKS on their short-lived bolt-on turbo kit for the SW20. GReddy has offered a 20G TD06 kit for years, and the TD06 has become an industry

Another advantage to buying a kit from KO Racing, ATS Racing or AutoLab is not so much what you do get, but what you don't. The problem with piecing together a turbo kit of your own, buying lines from vendor X, an adapter from vendor Y and a turbocharger from vendor Z: a kit that doesn't fit even after you bend your clutch lines 2" out of the way (which wasn't smart, by the way—don't bend yours, buy a kit that fits). This is frustrating and a sad reality of the automotive aftermarket.

SW2X STAGED BUILDUPS

This prototype represents AutoLab's latest continuous revision of their T3/T4 3S-GTE turbo kit. Note the adapter sandwiched between the stock exhaust manifold and the turbocharger that allows for an external wastegate placement right off the manifold. Photo courtesy AutoLab.

The T3/T4 hybrid kit has become quite popular for its affordable performance and easy application-sizing. In the KO Street Brawler kit, the choice of trim sizes allows user-selected choice in peak horsepower and powerband characteristics. Ultimately, the T3/T4 Street Brawler turbocharger will deliver top-end performance similar to the TD06, but with improved spooling characteristics. Photo courtesy KO Racing.

stalwart for 2.0 liter turbocharger performance even now, years after it was first introduced. ATS Racing basically went back and redid what had been poorly done the first time, creating two very affordable, well thought-out and complete, vice-free, vendor-supported turbocharger kits. Both are well sized to the 3S-GTE's 2.0L airflow needs. The TD06 is a flexible turbocharger, at home on a road course or a drag strip, while less suited to autocross courses, mountain roads, or commuting.

ATS Racing TD05/TD06 Features and Specs:
- TD05 or TD06 turbocharger
- TiAL 38mm external wastegate
- Machined (not cast) turbo adapter with wastegate flange
- External "dump pipe"
- Mandrel-bent stainless steel downpipe
- Oil feed line with block adapter, oil drain line
- Water lines and fittings, gaskets
- Stainless steel bolts, nuts, etc.
- ATS 550cc ROM-tune
- Flow-tested Toyota 550cc fuel injectors (4)
- ATS Racing–modified fuel rail
- Walbro 255hp fuel pump
- 170°F thermostat

Note: 3" exhaust, aftermarket intake and boost controller are highly recommended. Dual-map option available to run race gas for up to 19 psi and 400 horsepower.

AutoLab T3/T4 Turbo Kit—Performance Sweet Spot: 275–325 whp. A 46-trim compressor T3/T4 will spool a few hundred rpm faster than a 50-trim. A 50-trim will spool slightly slower, but add 15–30 whp potential, future room for growth.

AutoLab is a small operation that is big on custom modifications and performance, and this turbo kit was designed with that mentality in mind. Like the kit from KO Racing, the AutoLab kit allows user-selected choice in peak horsepower and powerband characteristics. Aspects of this kit include a quality TiAL external wastegate to prevent boost creep, a brand-new Garret turbocharger (some "kits" have been known to feature used turbos to keep the price low) and beautiful welds to ensure performance and reliability. Similar in performance to the TD06.

AutoLab T3/T4 Features and Specs:
- Original bolt-on T3/T4 turbo option in the States for the SW20
- Allows OEM "hot" pipe and stock intake system to be used
- Adequate clearance between compressor and clutch slave cylinder (don't take this for granted)
- Brand-new 30-day warrantied Garrett T04E/T3 turbo, 360 thrust bearing and choice of trim and A/R size
- TiAL 38mm wastegate, choice of color and spring
- Oil-cooled only
- Available at extra charge: 8-gauge TIG-welded carbon steel manifold, coated in 2000F protectant, lifetime warranty against cracking
- 304SS TIG-welded downpipe with factory O_2 style bung
- 1.5" 304SS TIG-welded dump-tube, vents to

atmosphere; can be routed back into downpipe for additional charge
• All oil lines, fittings, adapters and hardware included for complete installation.

KO Racing T3/T4 Street Brawler Turbo Kit—Performance Sweet Spot: 46-trim 275–325 whp; 50-trim 300–350 whp. Note: A/R is a measure of the turbine housing; smaller A/Rs will generally yield faster spool, at the cost of top-end airflow and ultimate horsepower potential. On this kit, the larger 50-trim will ultimately yield more horsepower compared to a 46-trim, but with slightly slower spool. If you are only looking for 300 whp, the 46-trim is a better choice.

KO Racing is a small performance-oriented company that uses feedback from the MR2 community to continuously improve their product. Their accessibility, customer service, and general end-user friendliness set companies like KO Racing apart from the traditional "giants" in the automotive aftermarket.

The T3/T4 Street Brawler kit is for the enthusiast interested in the bottom end performance that can only be delivered on a 2.0L engine by proper turbocharger sizing.

KO Street Brawler Kit Features and Specs:
• Full 3" coated downpipe with V band clamp and integrated 1.75" wastegate dump tube (with its own flex section)
• Wideband O_2 Port and Plug
• Downpipe support bracket
• TiAL Sport 38mm external wastegate
• CNC machined, coated steel manifold adapter, with integral wastegate port/mount.
• -3AN oil feed line
• -12AN oil return line
• Oil line adapter fittings for the block and oil pan
• Heat shielding (for clutch slave lines)
• Assorted hoses for connecting the wastegate to a manifold pressure source
• Assorted hardware (nuts, bolts, washers, gaskets, zip ties, etc.)
• Garrett T3/T4 0.63A/R stage 3 turbine/TO4E-46 compressor or Garrett T3/T4 0.63 A/R stage 3 turbine/TO4E-50 compressor

Camshaft Upgrade, Adjustable Cam Gears

Though camshafts ultimately serve the real workhorse of any turbocharged engine, the turbocharger itself, a Stage III SW20 has the potential to benefit considerably from an increase in lift and/or duration provided by a new set of camshafts.

Camshafts are crucial when seeking to absolutely

Camshafts are complex and expensive modifications to make, and as always, a cost-benefit analysis should be done before making the jump. Still, past 350 whp you are leaving as much as 50 whp on the table by not replacing the stock camshafts and cam gears. Web camshafts are pictured.

fully optimize your 3S-GTE, for example pairing a larger turbocharger with a redline capable of taking full advantage of that turbocharger's ability to move air. To achieve that redline, extensive changes in the valvetrain must occur, above and beyond changes to camshaft specification. These changes include changes to valves, valve springs, retainers, and even cylinder head porting to enable or supplement such changes—this in addition to cam gears, and proper degreeing of camshafts upon installation.

This complexity and financial cost is why this upgrade is pushed down towards the end of Stage III, and for many owners could even be skipped at this point. Though you're really leaving a decent amount of horsepower on the table—perhaps as many as 35–50 whp even at relatively modest 300–400 whp power levels—there is always something to be said for simplicity. An internally stock 3S-GTE with a properly sized turbocharger and proper timing and fuel control will make a lot of SW20 owners very, very happy, but for those that want to leave no stone unturned, the time to replace camshafts is now.

Which camshafts you should choose is not an entirely clear-cut thing, however. Through tuning, cam gear adjustment, and cam degreeing, performance can be maximized: area under the curve, peak hp, and idle are all adjustable through such tuning.

Ultimately, HKS 264 camshafts (non hi-lift), and HKS adjustable cam gears will allow your turbocharger to be used more so than the stock camshafts every could, while still producing strong area under the curve.

HKS 272s may provide superior peak numbers, and could be a wiser choice if you've got a very large turbocharger—T67, T78, GT35, etc.—but this choice must be weighed against a variety of other setup factors, not the least of which is engine management, and the rpm capability of your engine.

Contacting camshaft manufacturers or those shops specializing in performance head work with the details of your buildup is always a good idea, though be sure to consider all advice against vendor bias and salesmanship. Observe, collect data and opinions, and make a decision.

Note: It is critical that aftermarket camshafts are degreed according to their manufacturer specifications, and adjusted with cam gears on a dyno.

Limited Slip Differential

Though Cusco manufactures a very nice Type MZ limited slip differential for the SW20, it's difficult to source in the United States, much less get technical or warranty support for, or local feedback on, and we've established how undesirable this is. A Kaaz 1.5 LSD is therefore the best choice for the majority of SW20 applications.

Lighter Wheels, Wider Tires

You'll need application-specific tires here—drag radials or high-performance all-seasons, as your climate and driving calls for. What are you doing with your MR2?

This setup will be versatile, adding more rubber to the rear than the front without going so wide as to numb the steering response, or drastically increase rotational mass (wide tires are heavy). It should require no rolling of the fender, depending on how low your MR2 sits.

Note: There is opportunity here to adjust the handling balance to your liking. A 205 tire up front with the same 235 rear, on paper anyway, would increase the tendency for understeer by reducing the amount of grip available under the front tires, just as a 245 in the rear with 205 or 215 in the front would as well. As with sway bar sizing, rather than outright sizing, instead focus on the relation of size changes from front to rear. It is not recommended to reduce the rear width below 225, or the front below 205.

Stage III Recommended Wheel/Tire Setup:
- Front: 215/40/17 +38 offset, 17x7
- Rear: 235/40/17 +38 offset, 17x8

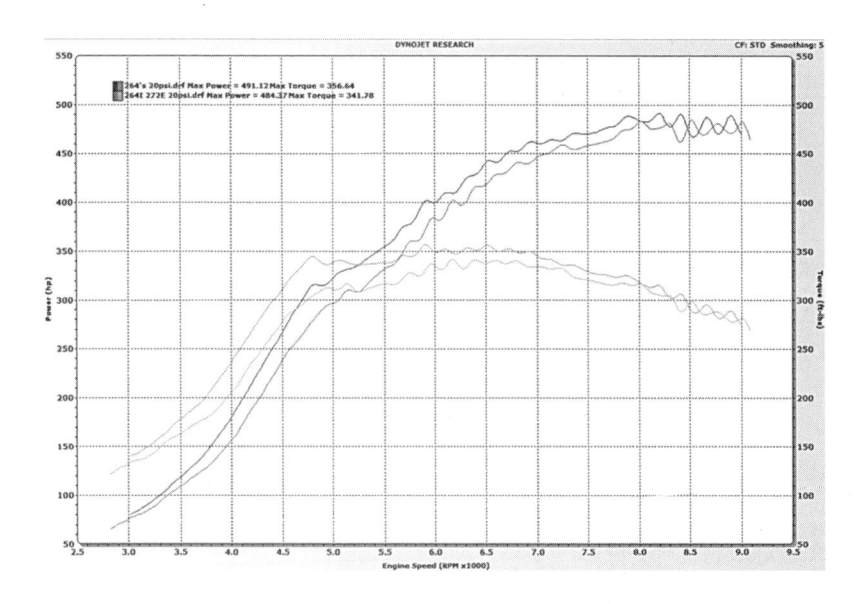

This dyno chart overlays the performance of a pair of HKS 264s, with a staggered setup, an HKS 264 intake camshaft and a 272 exhaust camshaft. Note the significant dip in horsepower around 5250 rpm for the 264I/272E combo, and the odd 30 whp drop just short of 8500 rpm for that combo as well. On this engine, at this power level, clearly the 264s were a better choice. On others, the 272 Exhaust camshaft could provide an upper rpm exhaust scavenging benefit. 264I/264E: 491 whp, 357 lbs-ft.; 264I/272E: 484 whp, 342 lbs-ft. Image courtesy of KO Racing.

A limited slip differential really shines on a road course in a 300+ whp SW20, allowing earlier application of throttle, and more predictable behavior afterward, helping improve exit speeds out of corners, and thus speeds down straights. Photo courtesy Vespremi/Crites/Smith.

TOYOTA MR2 PERFORMANCE

Obviously a unique, drag-specific tire width and tire compound is necessary to properly launch a high-horsepower MR2. A variety of factors input into selecting the right tire—climate, horsepower, driver preference for breakaway characteristics, and so on. The idea here in Stage III is that you are choosing a more narrow window of performance, moving away from universal parts, and instead choosing a wheel and tire package that is designed to do what it is you do with your SW20. Photo courtesy Condon/Fazio.

Proper offset is important, as it basically determines the flushness of the wheel with the fender. Improper offset can not only cause rubbing (and damage) of the tire, but it can just look plain bad too. This SW20 offset is picture perfect front and rear. Photo courtesy Michael Tedone.

Running up and down mountains or back country roads with pace requires a more forgiving wheel and tire setup than road courses or drag strips. Rather than choosing the "best wheels and tires," choosing the right wheel and tire package for your application is the correct approach. Photo Bryan Heitkotter.

ATS Racing Chrome Moly Inner CV Joints

We've discussed these previously. If you're drag racing, these became a must as early as late Stage II. At this point, there is no sense in having a stud engine and turbo setup rotating a fragile axle setup.

Upgraded Intake Manifold, Phenolic Manifold Spacer

Helping usher the now-larger amounts of air from your intercooler to your cylinder head, and equalize airflow across the cylinders, an intake manifold is now a worthy modification.

As discussed in Chapter 3, the Extrude honing process is more or less faultless except for the labor to remove and replace the manifold. For the majority of modifying SW20 owners—those who stay below 350 whp—an Extrude-honed stock intake manifold is probably the best choice. If you're paying a shop $80/hour to remove and replace the manifold, the cost of this modification—and many others like it—changes drastically.

In these cases, such modifications must be weighed for your specific circumstances. Replacing the intake manifold here is also an option, but it requires some sacrifice on your part. You may lose some degree of drivability—usually an aspect of low-rpm torque—with a short-runner/large-plenum intake manifold, though theoretically, a well-designed intake manifold could be developed to improve the performance throughout the powerband, from idle to redline. The challenge here is finding unbiased, apples-to-apples data—thus the recommendation to stick with an Extrude-honed factory manifold with a phenolic spacer until a point at which it is proven by extensive, unbiased, varied testing that an intake manifold exists that will undoubtedly improve the performance of the

146

SW2X STAGED BUILDUPS

This stock 3S-GTE intake manifold has been both Extrude-honed and ceramic coated as well to provide a thermal barrier. The Extrude-honing process not only seamlessly "ports" the manifold runners, increasing their air moving capacity, but also helps even out the airflow discrepancies that exist between runners on the stock intake manifold.

After doubling factory horsepower and boost levels, the tiny plenum of the factory Gen II 3S-GTE intake manifold cannot be ideal for the airflow requirements that now exist. The Ross Machine Racing 3S-GTE intake manifold dramatically increases plenum size, while reducing runner length. Once you've chosen a turbocharger that performs optimally beyond 5000 rpm, and high-rpm camshafts as well, narrowing your window of performance with an equally focused intake manifold makes sense. Photo courtesy Ryan Anderson.

3S-GTE in your engine bay. As of this book's publishing, that manifold—or supporting data, one or both—does not yet exist.

A 60mm throttle body, used on the stock 245 hp Gen III 3S-GTE, would also be helpful here, but good luck finding a bolt-on aftermarket 60mm throttle body, making something like this a custom prospect—again, time for a cost-benefit analysis. (And no, the Gen III throttle body will not fit.) If you're running all stock intercooler piping, stock intake manifold, and stock throttle-body inlet, the benefit of simply making one step in the intake tract larger may not be what you'd hope. Remember, if it isn't a restriction, or if other tangent restrictions exist—replacing it will yield compromised performance gains at best. In terms of true system bottlenecks, the throttle body is only a critical issue if you've addressed other related systems.

General Intake Manifold Recommendations:
- Stock up to 250 whp: stock intake manifold
- 250–350 whp: Extrude-honed stock intake manifold
- 350+ whp: RMR or custom intake manifold

Note: Any significant changes to the airflow of the engine—replacing the factory intake manifold for example—should be coupled with requisite changes to the engine management tuning.

Coil-Over Suspension Upgrade

Though a set of Illuminas or Koni Yellows can work extremely well with the relatively soft stock suspension springs, making for a decent ride and surprisingly potent handling, but with this much engine performance, more is called for from the suspension. In this case, the ultra-lightweight KW variant 3 coil-over is an excellent choice, but there are many. Other options include Ohlins and JIC.

Optional: Stroker Kit

Added displacement is a wonderful thing; so is torque, and an ATS stroker block is a good way to get it.

As always, consider the context of you and your MR2 when evaluating potential modifications. What's right for Joe may not be right for Bob. The reason a stroker kit is called "optional" here is because of the general complexity and cost associated with its implementation. It's expensive—and silly—to tear down a properly functioning 3S-GTE looking for another 0.2 liter of displacement. There are less expensive and more potent gains to be had elsewhere in most cases.

If, however, you are rebuilding your engine

Toyota MR2 Performance

> ## BRAKE SWAP
> ## 1991–1992 SW20 FOR 1993–1995 SW20 OEM BRAKES
>
> ### SW20 OEM Stock Brake Sizes
> This applies only the early 1991–1992 SW20, replacing calipers and brake discs with the larger pieces from the revised 1993–1993 SW20.
>
> Though larger rotors are also heavier, and won't necessarily shorten stopping distances, their slightly increased thermal capacity can increase their ability to resist fade, supplementing the Stage I brake system modifications we've already performed.
>
> **1991–1992**
> (Rotor Height x Thickness)
> Front: 258mm x 25mm
> Rear: 263mm x 16mm
>
> **1993–1995**
> Front: 275mm x 30mm
> Rear: 281mm x 22mm

Turbocharger selection is critical. Get it right, and your MR2 is a flexible, potent performer at home on a road course, as well as on the street, and even capable of 11- and 12-second quarter-mile passes. Get it wrong, and you've got an expensive one-trick pony. Photo courtesy Vespremi/Crites/Smith.

At this point in the SW20's development, all of the major weaknesses have been addressed. The car stops hard and accelerates harder; it's also adjustable in the suspension and engine management, and has a solid foundation upon which performance can be increased. Photo courtesy Vepremi/Crites/Smith.

already, it becomes a much more cost-effective modification, and the ATS Racing stroker kit is the way to go.

ATS Stroker Kit Features and Specs:
- 3S-GTE cylinder block bored, checked, decked as necessary
- 5S-FE crankshaft machined to accept 3SGTE connecting rods. Cryogenically treated for strength and hardness.
- Reconditioned 3S-GTE rods
- Wiseco forged over-bore pistons with rings, ceramic coated piston tops with heat barrier, ceramic coated piston sides with dry film lubricant
- Clevite 77 bearings
- Options: Shot-peened stock rods, Eagle Rods, ARP Rod Bolts, others by request.

A Stage III MR2 Turbo can be whatever you want it to be: tough runner, road course athlete, drag strip terror, a trailer queen or docile-yet-potent street car. The technology inherent in a turbocharged car, especially one now owning fully controllable engine management, allows the MR2 Turbo to be malleable, with shoulders broad enough that, with the right inputs, parts and adjustments, it can become exactly what you envision. That potential, though, can go in either direction, and so greed must give way to careful planning and patience when developing any platform, especially one as sensitive and temperamental as the MR2 Turbo.

DEALING WITH A BROKEN 3S-GTE
Turbocharged engines are wonderfully addicting things. Not only do they offer a rush of torque that begs just one more wide-open third gear pull to redline, they also readily produce extraordinary amounts of torque and horsepower when properly modified.

Mix these tendencies—addicting runs to redline, obtainable significant increases in power production, and engines with approaching twenty years and 150,000 miles on them, and acceleration is not the only thing that starts to happen quickly.

Too much timing. Too much boost. Not enough fuel. Exhaust gas temperatures too high. Broken

SW2X STAGED BUILDUPS

A 400 WHP 3S-GTE PROFILE

The followng setup, assembled by Mark Sink and Bryan Moore, was dyno'd at a modest 17 psi, producing 346 hp and 303 ft-lbs. of torque, incredibly good considering the very mild boost pressure. Not only is this a very potent blend of parts, with a larger turbocharger this setup could produce considerably more power, as budget and application changes.

Engine
- 3S-GTE
- Cunningham forged 5S-FE rods
- JE forged stroker pistons, with new rings and bearings, (1.50 rod-to-stroke ratio)
- 3S-GTE block
- ChrisK "street-port" head
- 1mm oversized intake and exhaust valves
- Ferrea valve springs and retainers
- Shimless bucket conversion
- HKS 272 intake/264 exhaust cams at 108 intake and 112 exhaust
- KO Racing Street Brawler T3/T04E turbo kit (50 trim, .63AR stage 3 turbine)
- KO 3" TKO exhaust
- K&N filter
- AutoLab Spearco IC
- Turbohoses 4" intake
- Hux fuel rail
- 850cc injectors
- SX regulator (set at 50psi with no vacuum)
- Walbro GSS-341 fuel pump

Electronics
- GReddy Profec B spec II
- Hydra Nemesis
- MSD 6A with blaster 3 coil

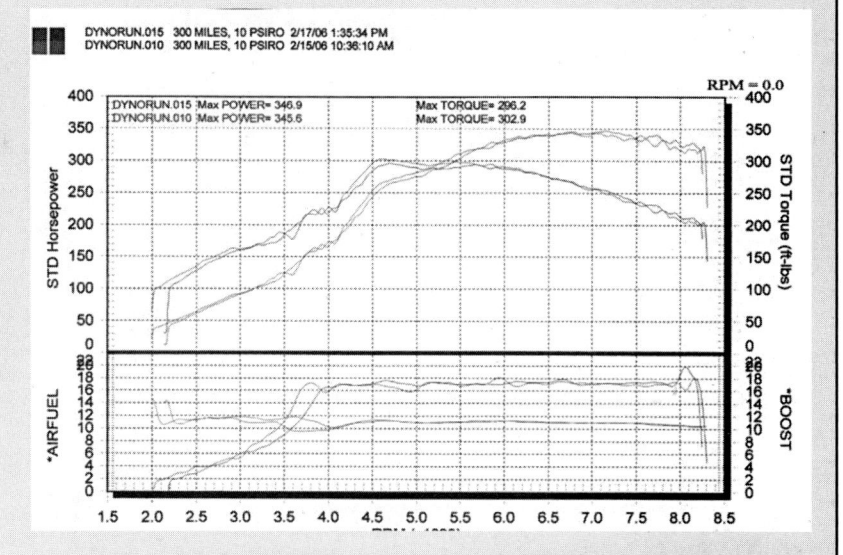

By 3300 rpm, the performance of this modified 3S-GTE has eclipsed stock peak torque and has nearly doubled it by 1300 rpm later. If you look at the bottom portion of the graph, you'll see boost and air/fuel ratio information. Consistent sub 12.0:1 air/fuel ratios and a very sane 17 psi, with full boost available at 3750 rpm thanks to a turbocharger well-suited to this application, and the extra 0.2l of displacement of the stroker kit. Final results: 346 whp, 303 lbs-ft. Image courtesy Sink/Moore.

---

ringlands. Worn piston rings. Too much or too little of something, and now something is broken. What to do? There are dozens of questions here, all depending on your garage space, mechanical know-how, financial state, and long-term ambitions for the car.

Sell it? Pay a dealer (your life savings) to replace the engine with a factory-spec engine? Pay your local mechanic a little less to do the same? Rebuild or factory rebuild? The questions are just now beginning.

If you've got the garage space, a decent set of tools, and something else to drive for a while, removing and replacing the 3S-GTE in your SW20 is not impossible. We've discussed engine swaps, including the purchasing used 3S-GTEs from engine importers, the reality of engine rebuilds, and the possibility of dealer-sourced 3S-GTE short-blocks. Consider these factors, then consider the following guide for removing a 3S-GTE from an SW20 as a supplement to the BGB. Any guide offering information on doubling your factory horsepower without attention to the potential consequences of such action—and review and further advice on how to address those possible consequences—is at best half-finished, and at worst irresponsible.

Removing a 3S-GTE from an SW20

Like any project, planning is essential. In addition to bagging and tagging hardware, it also helps to take digital photos of everything as it is removed. You'll thank me later.

When you get new parts from Toyota, take the time to identify where they go and what they do. Having a bunch of plastic parts bags with cryptic part numbers on them can be frustrating.

Track the part numbers you order—spreadsheets can help here. This can save you from duplicating the parts, especially when the project is spread out over a period of time. You never can tell when you'll get interrupted for several weeks, or even months, and searching for a part when you are not even certain you ordered it can be time consuming and frustrating.

I think it's false economy to reuse gaskets, seals, O-rings, belts, and hoses. Considering the amount of labor and inconvenience involved if they later fail, it just doesn't seem worth the small savings.

The BGB should be your source for torque ratings, but use common sense—as incredible a resource as the BGB is, it has errors as well. The BGB does include "standard" torque values for standard-size bolts, and these are useful as general guidelines. For example, if the BGB advises you to use 60 lbs-ft. on a 10mm nut, find a confirming source. It's likely you'll end up with a snapped stud or, worse, a stripped hole.

These pages have a lot of photos. That's really the whole point in creating this guide. Owner's manuals are often lacking photos. I use the photos as a reminder to myself. I often get interrupted during a project and can't return to it for days, even weeks. The photos can prove invaluable when trying to remember what I've done.

Ground Rules

While much, if not all, of this work can be done by a single person, working alone carries risks with it, so get some help when the more dangerous operations like lifting the car, removing the suspension components, and engine/trans removal are being performed.

I recommend removing all but a couple gallons of fuel. Sometimes, a disconnected fuel line will cause fuel to begin siphoning out of the tank. If you think this will be only a quart or so, and have a only small container to catch the fuel, you could be in for a big surprise. I like to use a container that can hold all of the fuel in the tank, just to be safe.

When I refer to bolts as "14mm" or "17mm," I am referring to the size of the bolt head—the wrench size. If I need to refer to the actual thread of the bolt, I will use different terminology, e.g., M10–1.25 x 45mm bolts.

Directionally, "Front" refers to the front of the car, "Rear" to the rear. "Left" refers to the left side of the car when standing behind it, and "Right" refers to the opposite side.

It's usually more efficient to break loose the mounting bolts for a component, then go back and remove them. Some of these bolts are very tight, and you'll find that loosening tight bolts when the other bolts have already been removed can be a real pain.

Make sure you have a plan to identify parts as you remove them. I used a system of small boxes and small parts bags. Label the bag and the box at the same time the part is removed, as specifically as possible. McMaster-Carr is an excellent source for small parts bags.

Tools/Materials Needed

A standard collection of mechanics tools (wrenches, sockets, screwdrivers, pliers, etc.). It will also help if you have the following equipment:
- Floor jack capable of lifting car very high—24" or more
- Tall, heavy-duty jack stands (24" height)
- Flexible plastic putty knife
- Large crowbar
- Sturdy mechanic's creeper for supporting the engine
- 32mm (1-1/4") 6-point socket for axle nuts
- Engine hoist or crane and lifting chain/hooks

The Procedure

1. Disconnect the negative battery cable.

2. Remove the engine lid and the side covers to provide access to the top of the engine:

3. Remove the strut braces by removing the 14mm bolts on the forward firewall, and the 14mm nuts on the tops of the strut mounts.

4. Disconnect the breather hose. It should come apart easily, as it is slippery with oil inside. The air intake hose may be stuck to the AFM and elbow. If you have to use a tool to "encourage" it to come off, avoid tools that would scratch the metal or tear the hose.

SW2X Staged Buildups

Removing the engine lid is smart anytime you're going to spend an extended amount of time in the engine bay. It comes out easily, and being fiberglass on the MR2 Turbo, is very light.

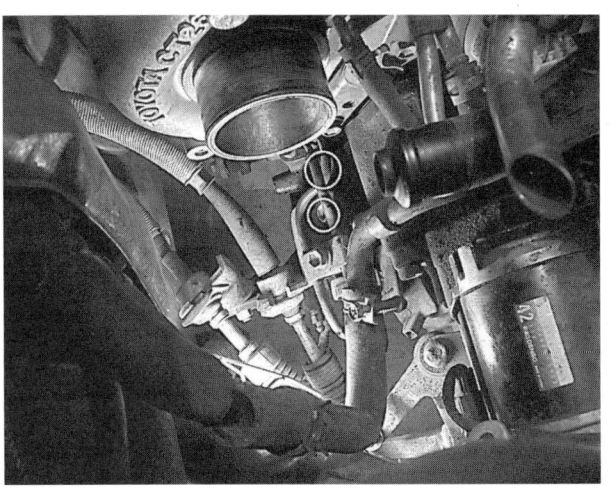

One of the bolts connecting the clutch slave cylinder to the transaxle is hidden beneath a couple of hoses.

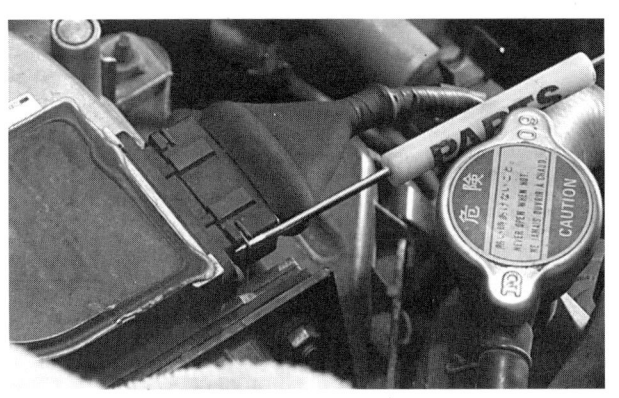

Loosen the clamps on the "bellows" hose that connects the AFM to the turbo and valve cover breather hose by addressing the hardware circled.

The AFM plug is attached with a spring clip and is easily removed with a small, flat-bladed screwdriver. Do not remove the Phillips screws.

5. Loosen the clamp that attaches the elbow to the turbo compressor, and remove the two smaller hoses that enter the elbow from the side. Remove the elbow and set it aside for now.

6. Reach down onto the transaxle, and locate two 12mm bolts that attach the clutch slave cylinder to the transaxle case, and remove both. Note that they are different lengths (the shorter one nearer the right side of the engine), so mark them after removal.

7. Replace the turbo compressor intake elbow to provide protection for the exposed compressor wheel. Stick a clean rag in the elbow to prevent anything from falling into the turbo.

8. Detach the electrical plug that connects to the AFM, but do not touch the Phillips screws!

9. Remove the clips on the air cleaner. Once the lid is removed, lift out the filter, and remove the 3 bolts in the bottom of the air filter housing. If you still have the stock ductwork, it may require some creative angling to get it out of that tight spot.

10. Loosen and remove the clamps holding onto the intercooler hoses, which Toyota calls the No.1 and No.2 intake air connectors. Then remove the hoses themselves. Inspect these and all other hoses for integrity. If they don't appear to be in tip-top condition, replace them.

11. After removing the cruise control cover, detach the intercooler fan connector and AC compressor clutch connectors.

TOYOTA MR2 PERFORMANCE

More fun with aged connectors and potentially brittle wires. The MAP sensor is used by the ECU as a boost pressure sensor, giving information to the factory boost gauge, and for fuel-cut purposes, unlike the Gen III 3S-GTE, which uses a MAP sensor to calculate airflow.

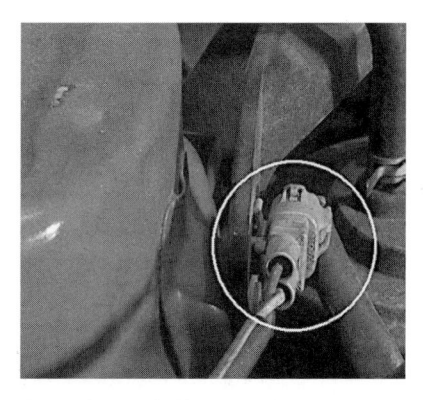

The left rear ABS sensor is shown here, next to the charcoal canister, which is shown with one of its hoses removed.

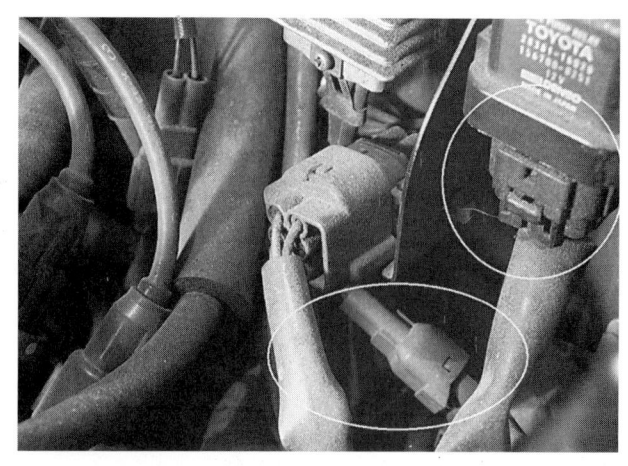

You only need to detach the connectors to the fuel pump resistor pack and the fuel pump relay, circled.

secures the fuel pump relay, the fuel pump resistor pack, the solenoid (fuel injector) resistor pack, and the Idle Air Control Valve (IACV). You only need to detach the connectors to the fuel pump resistor pack and the fuel pump relay.

24. The relays and resistor packs are held to a bracket on the firewall by two 10mm bolts. Remove the bracket, and lay the parts out of the way.

25. On the bottom of the IACV, there are two hoses. Trace the hose that leads to the back of the intake manifold. Loosen the clamp on the intake manifold nipple, and slide the hose off. This will give you enough slack to move the IACV out of harm's way.

26. Go the ECU in the trunk. Remove the trunk liner, and view the ECU. Remove the middle and left-side connectors by pressing in on the locking tab and gently pulling the plugs free. Because there are so many contacts, you'll need to wiggle the connector slightly while pulling.

27. Detach the white connector on the driver's side of the ECU by pressing in on the locking tab, the pulling the connector apart. Slide the captive end of the connector out of its holder.

28. Detach the starter relay and intercooler fan connectors, both to the right of the ECU. There's a cable clamp that needs to be removed as well. You

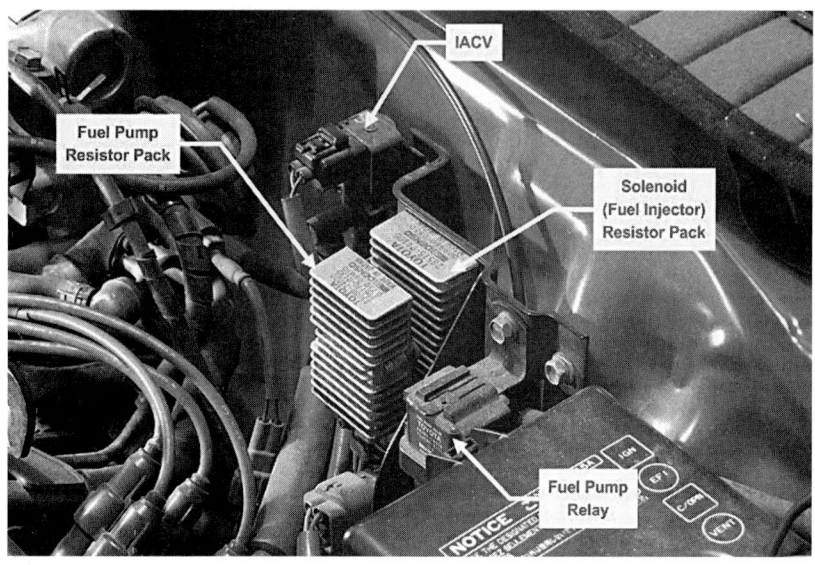

The fuel injector resistor packs are necessary to drive the stock low-impedance injectors of the 3S-GTE.

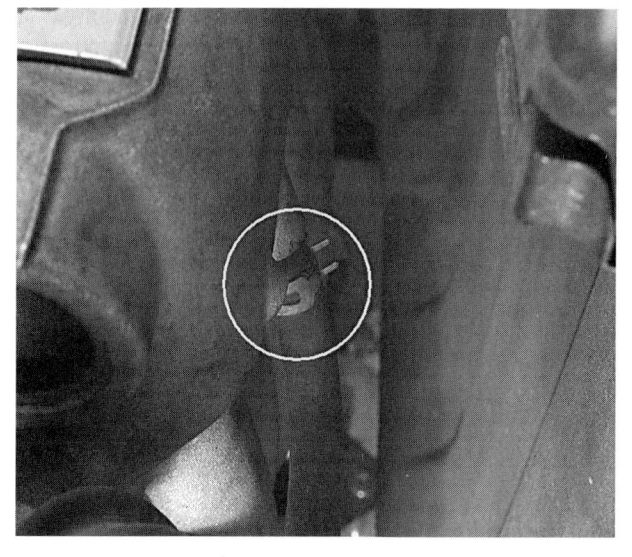

Removing the vacuum hose here is preferable to the working with the plastic IACV. It's a pricey part to replace if you damage it.

Remove the middle and left-side connectors of the ECU by pressing in on the locking tab and gently pulling the plugs free. Leave the right section attached.

Once everything has been unplugged and the grommet has been pushed through the rear firewall, the wiring loom is ready to be pulled through. Watch the potentially sharp edge of the hole itself.

need to pry it apart with a small screwdriver, but it's a real gem of design. Be patient with it.

29. Detach the connector and wire running from the distributor cap to the ignition coil.

30. Remove the cover from the fuse box in the left rear corner of the engine bay, and detach the two white connectors.

31. The main power cables are attached with two 10mm nuts. Carefully pull the red rubber boot away from the studs, remove the nuts, and detach the top cable assembly from the studs.

32. The fuse box is actually made up of several pieces—the top cover, the middle body, the bottom cover, and a harness retainer. You need to separate all of these pieces. First, detach the box from the car body by the various brackets and bolts holding it on, which requires several obvious steps. You'll note that with the fuse box free of the car body you've only gained a little slack, and there is not much freedom to move it around. The bottom cover is still held on with four locking tabs. Pry the locking tabs away from the body of the fuse box—it won't simply pull off.

33. The wiring harness that comes up through the left engine side of the fuse box must be detached from the box. The harness is taped to a mounting sleeve that slides down into the side of the fuse box, so removing it means sliding it back up and away from the box. Unfortunately, this is made more difficult by the position of the locking tabs, which Toyota engineers thoughtfully placed under the box. There are two tabs, one on each side of the sliding sleeve, and they must be pried away from

Detach the connector from the cap by prying the locking tab out away from the cap, and pulling on the connector (not the wire). You will probably need to wiggle the connector a bit to free it.

The two white locking tabs can be removed by pressing the locking tab towards the housing, then pulling the connector. This is a good time to be certain you've disconnected the battery.

the sleeve. Basically, you are working blind. Pull the harness up, and the sleeve should come free.

34. The fuse box is almost free! With the harness free, the bottom cover can be pushed down and away from the body of the fuse box. All that remains is to detach a cable clamp from the inside of the cover. Squeeze the little locking tabs together and push the clamp through the cover. The cover can now be removed. At this point, it makes sense to secure all of the loose components to either the engine or the body, to prevent them from get tangled when the engine is lowered out of the car. Zip ties work well here.

35. With the fuse box out of the way, you have access to a bracket that mounts the igniter, coil, and a condenser for the radio. There are two 10mm bolts securing the assembly to the rear firewall.

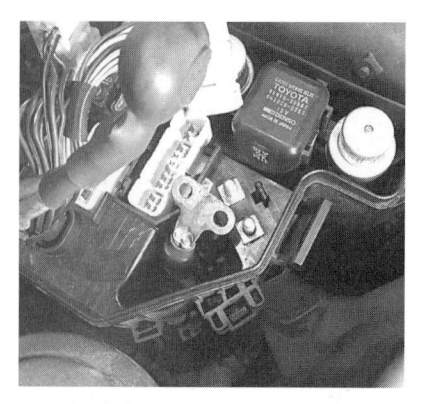

Toyota MR2 Performance

The case of the fuse box has a locking tab that secures the metal tab on the cable end. Pry out, away from the cable, and the metal tab should spring free, enabling you to remove the cable from the studs.

If you maintain upward pressure on the harness while trying to release the locking tabs, you should be able to free one side, then the other, allowing the harness to be pulled up, and free.

If you don't want to destroy the cover on the starter cable connector, you'll need to release the locking tab on the "lid." The tab is totally obscured, on the inside of the cover, and it needs to be pried out away from the housing to release

You can remove the plastic cover if necessary to access the two 10mm bolts, but it's not required.

The engine bay is really opening up now. Pay close attention to small invisible items like ground wires, other still-attached connectors, and even your exposed paint. Attention to detail pays huge dividends in large mechanical jobs like this.

You'll need to remove the bolt closest to the left side of the car. Rotate the unit up from the left side and it should slide out from the remaining bolt. Remove the coil wire to protect it from damage, then secure the assembly itself to the engine. Separate any grounds that you see dangling at this point.

36. Disconnect the starter cable (be careful with the plastic lid/housing).

37. The main ground cable is now accessible—it's black, and made of stouter gauge than most of the wires in the engine bay. It needs to be removed via its 12mm bolt, and laid out of the way so that it doesn't catch during the imminent engine drop.

38. There are two engine support brackets very close to the ground strap. The engine mounting stay is an angled tube that connects the left hand engine isolator to the transaxle case. The lateral control rod mounts between the isolator and the body. Remove all three bolts, then remove the engine mounting stay and set it aside.

39. Locate the fuel line near the firewall, and loosen its clamp. Slide the hose off the steel line. I recommend plugging both sides of the line.

40. Reach into the cabin and move the heater temperature lever to the full HOT position. This opens the heater core valve. Loosen the coolant filler cap all the way, and just leave it sitting loosely on the filler neck. If you only loosen it a little, the vacuum created in the system when you open the drain plugs will pull the cap back onto the neck and prevent proper drainage.

41. Break loose the lug nuts on the rear wheels, then jack up the car high enough to enable you to get all the way under it. SUPPORT IT SECURELY BEFORE CLIMBING BENEATH THE CAR. Four properly set jack stands should work; be sure that the car is relatively level. Not only is this safer, but the cooling system will probably not drain properly if only the rear of the car is raised. Remove the rear wheels.

SW2X STAGED BUILDUPS

You only have to remove the one bolt on the lateral control rod, and you can rotate it up out of the way.

To drain the transaxle, you'll need a 24mm box wrench or socket to loosen the drain plug.

The various splash guards come off next. These are often in disrepair, usually because missing hardware holding them on from inattentive previous owners. In their defense, they are held in places by fragile, troublesome plastic clips that are a pain to remove, and break easily.

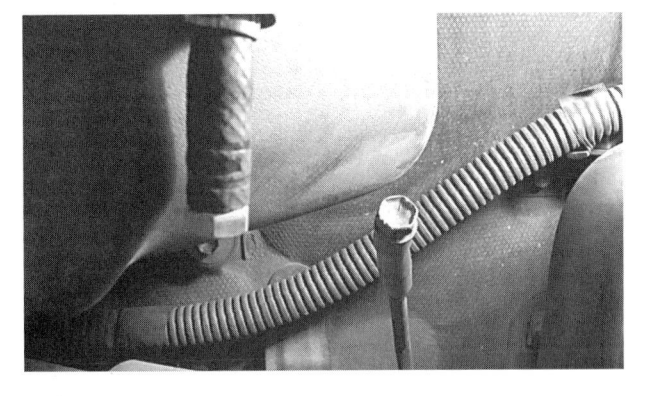

Plugging the line keeps any gasoline and fumes contained within the fuel system, especially important when working in a garage.

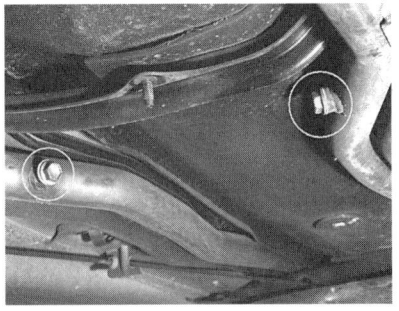

Position a drain pan with a capacity of several gallons under one of the radiator drain plugs. Loosen it with a 12mm box wrench or socket. Be careful as the coolant will shoot out quite a ways before the pressure drops. If you're changing your coolant or converting to NPG+ coolant, these are just two of the drain valves you'll need to use.

The clamps holding the hose to the coolant filler are easy to remove at this angle. Take note of the condition of the rubber beneath the clamps, and be sure they are in excellent condition. If not, they should be replaced without hesitation.

42. Drain the engine oil. While this is not absolutely necessary, it certainly is recommended. You'll need a 14mm box wrench or socket to loosen the drain plug.

43. Drain the transaxle.

44. There are several plastic splash pans under the central tunnel, and you'll remove three of them, working forward from the rearmost one. Toyota calls the two rear pans "Engine Under Covers," although they don't actually reach under the engine. The rearmost pan is attached with four 10mm machine screws and one 10mm sheet metal screw. Remove the fuel tank protectors at this time as well.

45. Drain the coolant at the two drain valves located on the pipes running from the engine, to the radiator at the front of the car. Leave the pan in place, as draining coolant likes to keep you guessing.

46. Loosen the clamps that secure the coolant filler hose to the coolant outlet manifold and filler neck. Twist the hose gently, and remove it, and do the same with the other coolant hoses in the engine bay—heater core hoses, and those entering the thermostat housing—as well. Be sure to zip-tie any hoses that may interfere with the engine drop out of the way.

Note: Clean any old gasket material completely off the coolant pipe drain bungs, and make sure to use a new gasket or sealing washer before reinstalling the drain plugs.

Toyota MR2 Performance

There are two heater hoses that go into the firewall, and those entering the thermostat housing as well.

If the crowbar isn't long enough to reach the ground, shim it up with a thick piece of wood, like this 4" x 6" timber.

47. Now it's time to jack the car up high enough to slide the engine out from underneath. Because of the height you need to reach (about 30" of clearance at the rear fascia), safety is a big concern. Obviously, you'll need heavy-duty jack stands that can extend to about 24".

The clearance from the garage floor to the lowest edge of the rocker panel is 23.5" at the rear, and maybe 2" lower in front. The clearance to the edge of the lower rear fascia panel is 30.5". Since the engine will be rolled out the back, this is the crucial dimension. You can set the engine on a creeper if you're careful, which is very low. If you use a taller dolly, you'll need to raise the car to compensate.

Note: Watch for late-to-spill coolant as you manipulate the car on the jack and jack stands.

48. Remove the cotter pin from the end of the rear axle shaft, and remove the cage that houses the nut.

49. Install lug nuts on three consecutive studs. Next, find yourself a big crowbar. If you don't have any air tools, you may need a long breaker bar plus a long length of iron pipe. Once the nut is loose, you can remove it.

50. Since the suspension and driveshaft need to be removed, the best approach now is to break loose all of the mounting bolts. Most of these require quite a bit of muscle to loosen, and it's best to apply that

Getting the axle nut off is a challenge. A 1 1/4" 6-point socket fits if you cannot find the correct 32mm metric size.

Don't remove the nuts on the rear ball joints/hub just yet; break them loose so you can spin them off easily later. One bolt is hidden from view in this photo.

force before you start removing pieces. Loosen the two 17mm bolts that hold the ball joint to the hub, and the 17mm bolts to the left and above.

51. Loosen the 19mm bolt that secures the lower arm to the crossmember.

52. Loosen the 17mm through-bolt that secures the strut rod to the body, and to the right of the strut rod bolt is a 10mm bolt that secures a cable retainer for the parking brake cable. Remove, rather than loosen, that bolt.

53. Remove the parking cable retainer bracket secured by two 10mm bolts.

54. Remove the U-clip with a pair of needle-nose pliers, and the E-clip can be pried out with a small flat-bladed screwdriver. Now the brake line joint can be freed from the bracket.

Note: If your bracket does not have a slot, you

SW2X STAGED BUILDUPS

The lower arm is attached to the crossmember by a single, beefy 19mm bolt. Loosen it.

Cutting a notch into this bracket will enable the brake line to be freed from the bracket without having to disconnect the line for future use.

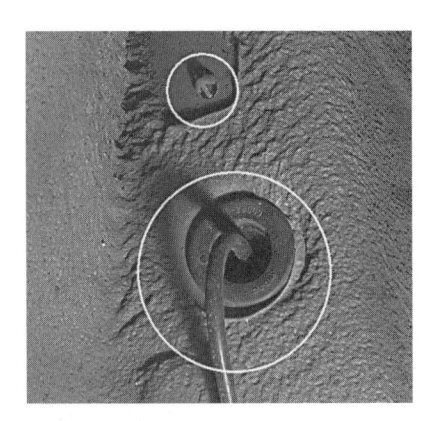

Once the ABS sensor has been removed, keep this cable out of harm's way by tying it to the hub.

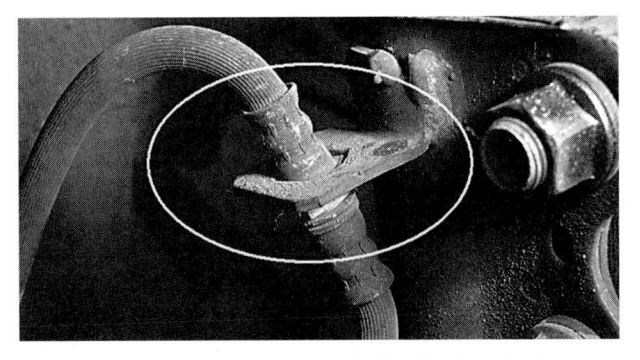

The brake line uses a support attached to the strut housing. However, rather than being simply notched, the bracket has a through hole. This means that in order to gain slack in the brake line you must disconnect the line, which in turn means that you must bleed the brakes after reassembly.

You can see the primary catalytic converter still in place in this photo, with an aftermarket exhaust from KO Racing bolted to it.

will need to disconnect the brake hose to provide enough slack in the cable to remove the caliper.

55. Inside the wheel housing, the cable to the ABS sensor can been seem coming through a rubber grommet. Squeeze the tabs together to free the retainer, and push the retainer through the hole to the engine compartment.

56. Remove the exhaust system, leaving the downpipe attached for now. If you've still got the stock exhaust, the car may need to be even higher in the rear to loosen the B-pipe.

57. Loosen the bolts circled on the and around the rear hub and trailing arm—as we've been doing, leave them in for now.

58. Loosen both 19mm through-bolts that secure the hub to the strut housing, and the 17mm bolt that attaches the lower suspension arm to the hub,

The circled bolts still need to be loosened, and they've all got considerable torque in them. Considering that your MR2 may be on jackstands, use caution while cranking on them no matter how secure the car seems.

Toyota MR2 Performance

The 19mm through-bolts securing the hub to the strut housing need both a ratchet and a wrench.

A 14mm box wrench and 5mm Allen wrench, preferably mounted in a socket, are necessary to remove the lower attachment point of the stabilizing link.

Pivot the end of the lower suspension arm up and tie it off to the sway bar to prevent damaging it when you drop the crossmember.

Now we can remove the bolts we had only previous loosened, ready to actually begin removing these heavy objects from the car.

The right rear lower suspension arm removed.

and the lower attachment point of the stabilizing link as well.

59. Now it's time to start removing components. Remove the two 17mm bolts that secure the ball joint to the hub. Also, remove the 19mm bolt that secures the lower arm to the crossmember, and the 17mm through-bolt that secures one end of the strut body to the car.

Note: Be sure to support the strut rod and you remove the 17mm through-bolt securing the strut body to the car body.

Make sure you are holding on to the strut rod and lower arm, as they will fall free when everything is loose. You'll probably need to slip a pry bar into the pocket in the body to pry out the end of the strut rod.

60. The brake caliper is next. You'll need some wire, rope, or long strong cable ties to suspend the assembly and avoid stressing the hydraulic line. Make sure these are within reach before removing the caliper. Remove the two 17mm bolts from the back of the caliper, then slide the caliper off the rotor. Tie the assembly out of the way, taking care to avoid any stress on the brake line.

61. Remove the 14mm nut that secures the lower point of the stabilizing link, and disconnect the link. Remove the 17mm bolt from the lower suspension arm, and remove the arm from the hub as well.

62. Check the hub to verify that only the two strut housing bolts are securing the hub. Remove the lower bolt first. You may need to use a drift pin to drive it out, since the weight of the hub will keep it tight. With the lower bolt removed, slowly drive out the top bolt while you hold up the hub. If you don't, the hub will drop when the bolt is finally removed, and you damage the driveshaft or the hub in the process.

63. We need to remove the support bearing from the driveshaft, however dowel pins on the block locate the support, and there is not enough slack in the driveshaft assembly to permit the bearing support to come free until the shaft is pulled free.
There's a snap ring on the right side of the bearing, and it should come out easily with a pair of pliers. When the snap ring is out, loosen the 14mm bolt on the top of the support. The driveshaft should now pull cleanly out of the transaxle case.

64. Remove opposite driveshaft.
Note: Removing the opposite driveshaft is nearly identical proced-urally, with the exception of the left shaft lacking a support bearing.

65. Remove the rear engine mount via the three 14mm bolts beneath it.

SW2X STAGED BUILDUPS

Once the top mounting bolt has been removed, you should be able to slide the hub off of the axle.

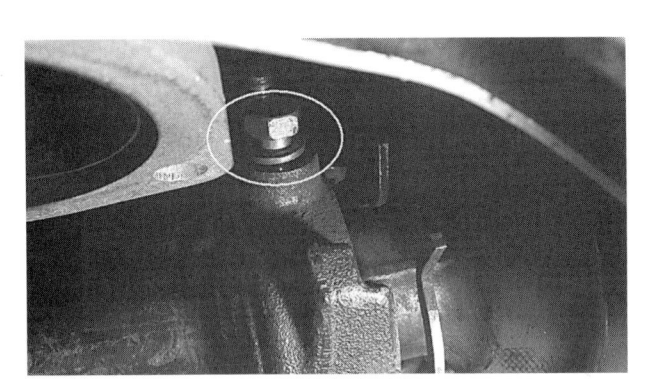

With the driveshafts removed, we are almost ready to begin removing engine mounts, and dropping the engine.

About 10" from the transaxle case, there's a support bearing on the driveshaft. There is a snap ring on the right side of the bearing that needs to be removed.

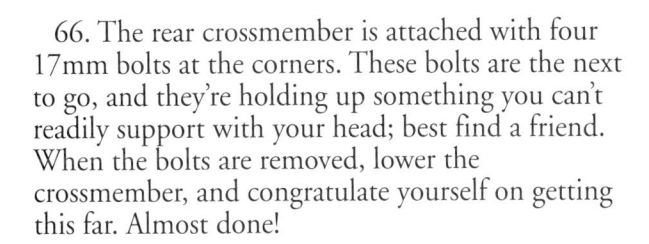

With the splash guards gone, the wheels removed, the axles on the ground, and the suspension pieces half-gone, things look a lot different now.

The rear engine mount is anchored by three 14mm bolts.

66. The rear crossmember is attached with four 17mm bolts at the corners. These bolts are the next to go, and they're holding up something you can't readily support with your head; best find a friend. When the bolts are removed, lower the crossmember, and congratulate yourself on getting this far. Almost done!

67. There are two stiffening supports that Toyota calls "Lower Suspension Supports," although they don't really attach to the suspension members. Remove these bolts, and the supports, and the A/C hose supports as well.

68. Locate the fuel filter, remove the 17mm banjo bolt on the top, and tie the hose end up to the engine. Remove the fuel filter assembly from the car. The A/C compressor also needs to be pulled off the engine block and tied off to the body, out of harm's way.

Grab a strong friend and have them support the crossmember as you loosen the bolts holding it on.

The 3S-GTE is almost ready to come out. Just a few minor details and checks remaining.

69. Now here's the last pain. You need to remove the remaining idler pulley bracket bolt. It's located just above the two large pulleys, on the right-hand side (timing-belt side) of the engine. It's a 12mm bolt and it's very hard to reach. You'll need to

Toyota MR2 Performance

The 14mm bolt securing one end of the right stiffener—the other end is under the A/C hoses. These can go now.

This 12mm bolt is very difficult to even see, much less reach. It must be removed to remove the idler pulley bracket.

When removing the fuel filter, have some (fuel safe) container to catch any fuel that leaks out. This photo shows the lower banjo bolt removed, which is fine if you are replacing the fuel filter while the engine is out anyway.

Here's one of the lower A/C compressor bolts, located just above the oil return hose that runs from the turbo to the side of the oil pan.

experiment with various lengths of sockets and extensions, as the intercooler is very close and vulnerable. Once you remove this bolt, you can remove the idler pulley bracket from the top of the engine. Do this before continuing.

70. Remove the lower A/C compressor bolts, located just above the oil return hose that runs from the turbo to the side of the oil pan. These are both 12mm bolt heads, and they are quite long.

You can remove the compressor away from the engine by carefully twisting and easing it down. It must be secured well, because it's heavy.

71. Move the engine hoist into place, and secure the engine to the chain. Once you have the engine hooked up, put tension on the chain by raising it until the engine's weight is on the hoist. This will make it easier and safer to remove the engine mounting bolts.

Note: Remember, you'll be lowering the engine about 24", so take some measurements and make sure you have enough clearance where the boom is closest to the car body. If you have your hoist positioned in the rear, make sure you can lower the hook at least 24" before the boom hits the rear spoiler. If you fail to measure this now, you might find that you need to raise the engine back up, secure it to the car, and add some length of chain. That's a lot of work that can be avoided by checking now. Measure twice, cut once.

72. One of the shifter cable ends is still attached to the shift arm. Remove the retaining pin on the lower shifter cable end and slide the cable end off the shaft of the shifting mechanism. Pry the U-clips off of the shifter cables, and separate the cables from the bracket, then move them out of the way of the transaxle.

73. There's a 17mm through-bolt that attaches the mounting bracket to the engine via a rubber isolator. Break this bolt loose, but do not remove it yet.

74. There are four 14mm bolts that secure the same mount to the body at the forward firewall. Break these four bolts loose, but do not remove them yet. Remove the large through-bolt, then remove the 14mm bolts and slide the mount off the transaxle bracket.

75. Two 14mm bolts secure the transaxle mounting bracket to the transaxle case. Remove them both, and then remove the bracket.

SW2X Staged Buildups

Have some rope, wire or strong cable ties handy to tie off the A/C compressor before you begin. You need to turn the compressor to face the pulley side down, then ease it by the various wires and brackets. Just be sure to keep it fully supported, then you can tie it off to the strut rod mounting bolt where the caliper is tied.

There are two U-clips that retain the shifter cables to the support bracket. These must be removed to free the cables.

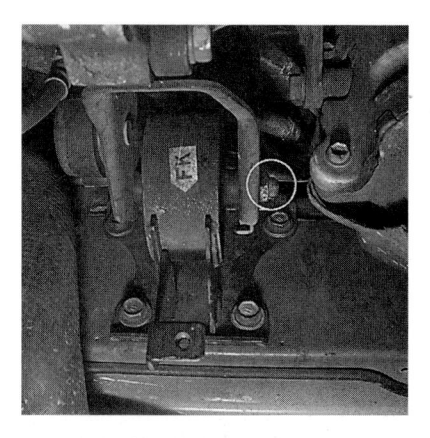

The front mount, like the rear mount, is really a torque control mount, and doesn't take much weight when the car is at rest.

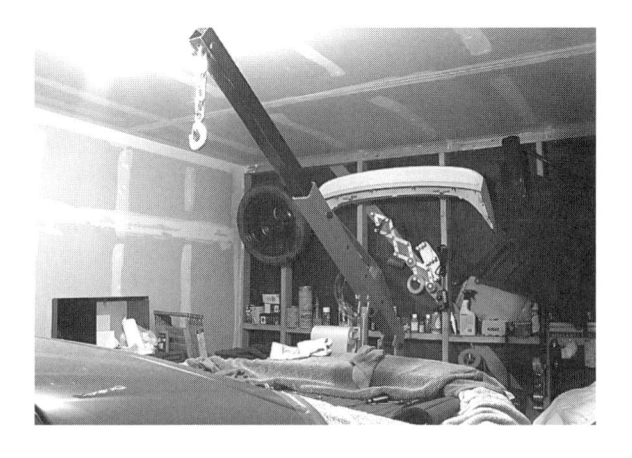

After breaking these four bolts loose, you'll need to remove the through-bolt. You may find that rocking the engine a bit helps to free the through-bolt if it gives you trouble.

After removing the two bolts securing the mounting bracket to the transaxle case, the only anchoring points left securing your 3S-GTE should be the left and right engine mounts.

If you don't own an engine hoist, or "cherry picker," oftentimes local equipment rental shops carry them, and they can often be rented for less than $50 per day.

76. With a little maneuvering, you should now be able to pull the clutch slave cylinder assembly down away from the transaxle, and tie it up away from the transaxle and engine.

Note: Take a moment now to verify that all brackets, hoses, and cables that attach to the engine have been disconnected.

77. From the top of the engine, look down in front of the timing belt cover and you'll see three 14mm bolts that secure the right-hand engine mounting plate. Remove these three bolts, as well as the mounting plate.

Note: Before removing any more bolts, double-check your hoist and chain to ensure that everything is secure. The weight of the engine and transaxle assembly should already be on the hoist, so the chain should feel extremely rigid. Also take one last opportunity to check the engine bay to be certain you didn't miss a connection, wire, or hose somewhere. They hide.

Also, when removing these final bolts, never place your hands or any part of your body in danger. Remember, when you remove a bolt with the weight of the engine on it, it's likely that the engine will lurch in one direction or another. Make sure you don't rest any part of your anatomy where it could be crushed.

78. On the left side of the engine, there's a 17mm through-bolt that runs through a rubber isolator. This mount supports the left side of the engine/transaxle. Remove the nut, then carefully remove the bolt.

79. On the right underside of the engine are two 14mm nuts that attach the engine to the isolator. Remove these two nuts.

Beside the timing belt cover, there is another engine mount. Once this has been removed, there is only one mount left. At this point, the weight of the engine will be shouldered by the engine hoist.

The final two 14mm bolts separating your 3S-GTE from freedom. Use caution getting at these two—do not place any part of your body beneath the engine to remove them!

The remaining through-bolts are not visible when viewed directly above, but they are accessible. When you remove them, you will probably find that there is still weight on it from the transaxle. It's difficult to get the hoist balanced exactly on the engine/trans assembly. However, by carefully adjusting the load on the hoist and by rocking the engine slightly, you should be able to remove the bolt.

Finished! Be cautious moving the engine. Not only can you knock the car off the jack stands, but you can damage the engine itself.

80. The engine can now be lowered onto a dolly, or whatever method you've devised for moving the engine once it's out. You might need to rock it a bit to slide it off the right-hand mounting studs. Be certain that it has found a "resting place" on some straight edge of its belly before you attempt to move it. Be careful not to damage the oil pan!

81. Be sure there are no leaking hoses or other moving fluids, and that all your ties and other makeshift arrangements survived the lowering of the engine.

Thanks to Dave Martin for contributing his time and effort, as well as his technical and photography skills in this write-up. His thoroughness is inspiring for all shade-tree mechanics and would-be pit crew wrenches everywhere, evidence that with a little planning, patience, and attention to detail, removing and replacing an engine is not beyond the reach of an MR2 owner of average mechanical skill.

SW2X STAGED BUILDUPS

AN INTERVIEW WITH STEPHEN GUNTER

Stephen Gunter is an experienced member of the SW20 tuning scene, and unlike many other folks who have moved on to other cars, has stuck with the MR2 through the years, and is a self-admitted Toyota addict. In his 20+ years of experience chairing Jamaica's Drag Racing association, rally racing and engine building, he has acquired knowledge that would benefit any MR2 owner.

You have participated in many forms of racing, including drag racing. What about the MR2 makes it suited to drag racing? Why isn't it more popular in drag racing?

The mid-engine configuration of the MR2 is what bests suits it to drag racing. Having the majority of the weight of the car over the drive wheels of a rear wheel drive car greatly helps traction when weight transfer takes place during the launch stage of a drag race.

It is not uncommon for a completely stock MR2 to record sub-2.0 second sixty-foot times, and this improves when better tires and suspension are added to the equation.

I believe the MR2 isn't more popular because it is looked at as an exotic of sorts, and because it doesn't have the traditional front-engine rear-drive setup. If you think about it, not many mid-engine production cars are popular in drag racing, although the fastest drag racing machines on the planet are mid-engine.

What are your thoughts on the intake manifold on the Gen II 3S-GTE? When does it need to be modified or replaced—and how or with what?

The Gen II 3S-GTE TVIS intake works well for the job Toyota designed it to do. If you left the engine alone, you'd never have a problem caused by the intake manifold. Overall, the intake design idea was good, since it promoted intake-charge velocity at low rpm for good engine response off-boost. From that point of view, it works. Of course, not many people leave engines stock these days, and that's how the flaws of the manifold were uncovered.

It really is difficult to say when the stock manifold "needs" to be modified or replaced, but most agree that the 300 whp mark is a good point. Of course, there are many MR2 owners that have made more that 300whp still using the stock intake manifold, but at that level you will notice a difference if you modify or replace just the manifold itself.

There are three modifications to the stock manifold that people can consider. The first is to put it through the Extrude hone process, which results in greater, and more importantly, evenly distributed airflow (hand-porting can have the same effect). The second would be to completely remove the plenum and replace it with a larger one, while retaining the stock runners. And the third is a combination of the first two: porting the stock runners as well as replacing the plenum. The "Surge Take" intake from JUN is the best known example of option number three.

Beyond those three, you have the option of an entirely different manifold design, usually involving a 4-runner design inside of the standard 8-runners of the stock manifold.

What causes Gen II 3S-GTEs to "pop?" Is it some design flaw involving the intake manifold, fuel rail or stock ECU settings, or is there anything at all, in fact, wrong with the Toyota's work?

Well, detonation is what causes them to pop, that part is pretty simple to understand. It only happens after people start modifying the engine and push things beyond what they were designed to do.

In strict technical terms, there is nothing wrong with Toyota's work, because the engine and electronics work very well when left alone to do what Toyota intended. Even when the modification process is started, the engine and electronics do a good very job of generating more power safely. It is only when the safe limits of the stock 440cc injectors are reached that bad things start to happen, and that is simply because now you're talking about making more power than the system was designed for. And that's an important thing to keep in mind when thinking about any engine: there is no one component to concentrate on; everything is part of a system. And once you push beyond the limit of any part of the system, it will cause a problem, which will probably not affect the overworked part itself. For example: going beyond the limit of the fuel injectors does not affect the injectors themselves: it affects the components in the combustion chamber that have to deal with the effects of a lean-mixture. Just remember to think of systems and their limits when approaching modifying an engine.

(continued on next page)

(continued from previous page)

If you were building an all-out race engine, which platform would you start with: Gen II 3S-GTE or Gen III?
Hmmm...Gen III.

What is the most common mistake people make when they start building their MR2s?
Unrealistic horsepower expectations from the stock components. An example is that people don't seem to want to hear that they shouldn't run the CT26 over 15 psi. I'd say that unrealistic expectations lead to the vast majority of stories of woe you hear about the MR2. Also, a lot of people forget that the MR2, the SW20, was built as a GT car, and not a straight line racer, and lose sight of the fact that it is a wonderfully balanced driving machine. Building a MR2 and concentrating on the engine is very short sighted, as energy spent further developing the suspension and brakes would significantly increase the fun factor of driving an MR2.

Chapter 7
Autocrossing Your MR2

Any generation of MR2 possesses considerable potential that can be applied in a variety of speed contests. Photo courtesy Vespremi/Crites/Smith.

So you've enhanced your MR2's abilities with an upgrade or two, and want to let it loose a bit. What the car "feels" like is absolutely crucial, but what is essential to any respectable project is some sort of data that tells you where you stand and where you can improve both your MR2 and your skills. What you need is a proving ground to let you know how you're doing, and what you and your MR2 might benefit from next. That is where autocross comes in.

SCCA AUTOCROSS

Across the nation, the SCCA (Sports Car Club of America) sanctions local motorsports events called autocrosses. An autocross is a relatively low-speed, high-intensity exercise in automotive performance. They are generally held in parking lots, with the course marked off by cones.

An autocross offers the average MR2 owner an opportunity to explore his own limits and the limits of his MR2, in a relatively safe environment. Many events have seasoned drivers willing to ride with beginners to offer pointers, and start them down the path of performance driving. Autocrosses can also be a good place to meet people with the same technical car interests who are often willing to share various methods of car setup and track tuning—from basics, such as chalking your tires or adjusting your seat/steering wheel position, to more complicated concepts such as boost levels and alignment properties. If you are lucky enough to find another MR2 owner out there who is willing to chat, all the better.

Autocross and Its Effects On Your MR2

Depending on your personality, beating up your MR2 on an autocross course might seem against your nature. After all, you love the car and have spent countless hours and dollars getting it into this condition, so why would you go and do a stupid thing like race it? The answer is, if you really do love your MR2, you'll take it racing. It wants to go.

MR2s were built to be driven hard. Even bone-stock they are very capable sports cars. With very few adjustments you can autocross your MR2 for many years for minimal cost and without shortening its lifespan.

Autocross is very hard on two areas: brakes and tires. During an autocross you basically spend 45–90 seconds at the very limit of your car's grip—in all four directions, being as you are (somewhat) sliding around on top of four very small contact patches of rubber. The fact that autocross is hard on tires should not be surprising. But tires are easily replaced.

Same with brake pads and rotors. The first ride I ever had on an autocross course was around 1994, appropriately in an SW20. It was a 1993 SW20 hardtop, a very rare bird indeed. I don't very much recall the weather that day, but I do remember the incredible braking force I felt, akin to the last 25 feet of a roller coaster when the driver hit the brakes. While it's true that forces are magnified significantly from the passenger seat, I still had never before ridden in such a fast car around such a tight course with such a capable driver, and it totally changed my perspective on what a car—and its brakes—were capable of. Again, pads and rotors are

If you want to be competitive, having an extra set of wheels to mount R-compound tires on is the way to go, even if they aren't an expensive set of Volk TE-37s from Rays Engineering. Brake pads are fair game in stock class autocrossing; replace the calipers with a set like that from AP Racing shown, be prepared to be moved up in class. Being a 4-bolt wheel, this is obviously an AW11 or ZZW30 wheel, as the SW2x featured a 5-bolt pattern. Photo courtesy David Jones.

READING LIST

We're going to focus our energy on offering basic autocross information and tips, mainly as they relate to the MR2. For more in-depth autocross information, your best bet is a book called *Secrets of Solo Racing*, by Henry Watts (ISBN:0962057312).

Though the book was published 1989, its wisdom is timeless. Watts explains the delicate intricacies of wiggling and wagging through those troublesome cones with the patience of a monk in a rocking chair, with literally every detail you might ever require in this dynamic, competitive activity. The book really is very good. The *Speed Secrets* series by Ross Bentley is another good recommendation (Book 1, ISBN:0760305188).

Be sure to use these simply as resources to supplement the best teacher and seat time on the track.

---

replaceable so don't be afraid to wear them out.

Ultimately Pro Solo–style launches will certainly wear your clutch faster, too, and is undoubtedly harder on your engine than interstate cruising. If you aren't heel-toeing and rev-matching on your downshifts, it can also wear your transmission as well, but remember, this isn't hot-lapping on a road course for 25 minute spurts. Autocross sessions are much briefer, albeit more intense, usually between 45 and 75 seconds.

In the end your MR2 is a sports car with high limits born to be driven near those limits. All generations of MR2 are extremely durable, and very affordable to run and race. My last full year of campaigning my SW20 at local events, the only item I had to replace other than tires was the windshield wiper fluid reservoir after it collided with a cone. Things happen, and stuff can break, but there is a marked difference between driving your car at its limits and abusing it. When you learn to do the former, the latter will no longer be a major concern.

Autocross Car Preparation

In the week leading up to the event, there are basic adjustments you can do to your car to increase the chances it will perform at peak levels on race day. What exactly this entails for you and your MR2 depends on the class you're running in, and its current condition. Assuming, like most folks, you're running in a stock class, or even one of the increasingly more popular street-tire classes, these adjustments are somewhat limited.

The Week before the Event—A basic "one week out" checklist to help you prepare for autocross.

1. *Verify your MR2 is OEM healthy.* Before you get to the event, check your car's vitals to be sure you're in perfect running order. You will be required to pass a basic technical inspection before you run. Be sure the battery is properly secured in using OEM or OEM-like hardware. Your MR2 should maintain proper operating temperature, should not smoke, possess a smooth idle and power delivery, normal brake operation, proper clutch feel and operation, and so on. If you wouldn't drive the car across the country, you don't need to autocross it.

2. *Inspect your tires.* Though you'll need to check your tire pressure again right before you run, checking it the week before will also allow you to inspect your tires for signs of chording, flat spotting, or other issues you may have to deal with.

3. *Remove loose and other unnecessary items.* This is for safety as well as performance.

4. *Consider fuel level vs. weight.* Depending on whether or not you drive your MR2 to the event, you want to try and arrive at the event with around a 1/2 tank of gas, but no less than 1/4 tank. You certainly don't want to be fuel-starved in the corners, but obviously too much gasoline brings with it a weight penalty—each gallon of gas weighs around 6 lbs. With this in mind, plan your gas

Autocrossing Your MR2

Depending on whether you drive or trailer your MR2 to the event, you want to try and arrive at the event with around a half tank of gas, but no less than a quarter tank. Gasoline does weigh around 6 lbs. per gallon, so the difference between a full and half tank could approach 50 lbs. Photo courtesy Bryan Heitkotter.

Before your first run, you need to remove excess weight and all other "non-essential items" anything from your MR2—basically anything not bolted down. This includes floor mats, mp3 players/CDs, as well as your spare tire and jack. Photo courtesy Helder Carreiro.

If you're going to run on tires other than what you drove to the event on, you'll need a jack. Keep in mind your ground clearance, though. If your MR2 is lowered, a standard jack may not fit. Photo courtesy Bryan Heitkotter.

consumption during the week so that you end up where you want to be when you are lining up for your first run Saturday or Sunday morning. I don't recommend that you drain the fuel.

5. *Preregister for the event, if available.* Many regions are now offering Internet preregistration, saving valuable on-site time. The less you have to do the morning of the autocross, the better.

6. *Gather supplies.* Stash as much of this as possible in an established location so you don't have to round it up before every event. Call it your autocross pile, and tell your spouse and/or kids to back off. See sidebar on page 170 for a suggested list.

High-Performance Tires

Being good to your tires should be among the first rules you learn in self-contained motorsports like autocrossing and road course hot-lapping. A tire cording basically means that it has shed enough of the tire's tread to reveal the cords—and means the tires are spent. This can happen very suddenly, with the tire's carcass basically failing and shredding into strands, or it can happen gradually. When a tire cords or you flat spot it, unless you've got a replacement with you, your day is usually done.

Flat spotting occurs when you lock up your brakes, and the tire is basically shaved on that spot, to the detriment of its shape. Softer compound of high-performance tires makes them especially vulnerable to this. One thing you can do to prevent this from occurring is make sure your brake system, including the ABS, is tuned and working perfectly. Make sure the fluid has been topped off, the pads and rotors are in good shape.

This is especially troublesome for AW11 owners, with no ABS and front tires that sometimes like to lock early. To avoid this is simple: avoid locking the tires by squeezing the brake pedal down until you find the locking threshold, the point in the pedal travel just before the brakes lock. This is also called threshold braking. If the tires don't lock, they won't flat spot either.

If you are driving soft compound tires on the street, also beware of gravel and other debris. The gummy tire compound of higher performance tires will pick up and spit out almost everything it rides over, peppering your MR2's paint with a road-debris potpourri all the way home.

Great care must also be taken to avoid road hazards of all sorts when driving on these tires, and driving in the rain isn't a good idea, either. Even if

AUTOCROSS SUPPLIES

What should you take with you to an autocross event? These events often last 6–8 hours, depending on how many attend events local to you and how well the event is coordinated. The following list is a good place to start.

Appropriate autocross supplies may include:
- Snell-standard helmet. Requirements may depend on your local region's regulations—not just any helmet will do; loaner helmets may be available at the event.
- Tire pressure gauge
- Shoe polish to chalk your tires, or in lieu of magnetic or vinyl numbers
- A quality torque wrench
- A hydraulic jack, though if you get good with a scissor-jack, they can be surprisingly effective
- Air for the tires—a compressor tank or accessories plug-in pumping device
- Basic tool set: metric ratchet set, wrenches, screwdrivers, extensions, etc.
- Duct tape, wire cutters, electrical tape, zip ties
- Sunscreen, sunglasses, poncho, windbreaker
- Water, food/cooler, paper towels, chairs
- AAA card
- Camera

Failing procurement of the entire list, at least bring the tool set, air compressor and shoe polish.

An SW20 drifting. Drifting is basically an exercise in grip control, or rather in controlling your lack of grip: You want to find the point where the available grip of the tires is nearly exceeded, and keep it exactly there, neither exacerbating nor repairing the slide. Like autocross or road course lapping, it is also obviously very hard on tires. Photo courtesy Sarah Mays.

The same soft, gummy compound that endows R-compound tires with such incredible grip also makes them especially vulnerable to flat spotting and cording, so check their condition after each run. Kumho Victoracers are pictured.

they are labeled "street-legal," extremely high-performance tires are generally best-suited to dry tarmac and little else—especially in an MR2. MR2 handling changes fast in the rain and snow, and riding around on a set of Hoosiers in the rain is tricky to say the least. Consider yourself warned.

Street vs. R-compound Tires—When you first start out autocrossing, simply running on your street tires is probably your best bet, assuming they're of at least decent quality. The morning of an autocross can be a hectic affair if you haven't planned properly, and is busy even if you have. There is last-minute car preparation, a tech inspection, course walking, and a driver's meeting all to be done before the first car is waved on the course. Running on street tires, rather than jacking up your car and swapping on another set of wheels, means one less thing to do before your first run, a positive thing for the beginning autocrosser.

More importantly, the lower grip provided by street tires also can also be far more educating. The immense grip afforded by R-compound tires can overwhelm you if you're not prepared for it. And while providing extraordinary grip, when they do let go, they usually do so in a hurry, in contrast to the noisy but helpful howl of street tires losing grip. The cost of R-compound tires mounted on a set of quality wheels can also quickly pass $1,000, something to consider if you're still deciding if autocross is your cup of tea.

That said, learning to autocross on R-compound tires has been successful for some, and it has its merits. Decide on your own, but consider the pros and cons of each method as you make your decision.

Other Issues with Soft Compound Tires—Maintaining appropriate tire pressure through consistent monitoring and adjustment is also important, not only for performance, but for overall tire condition and life as well. Running tires at their limits with too little air pressure can sacrifice the shape of the tire, which is important for not only grip, but its "adhesion" to the wheel itself.

AUTOCROSSING YOUR MR2

The increased grip offered by R-compound and other similar tires is substantial compared to your average performance tire, but the added traction can mask many of the "mini-transitions" a car makes while on course. These episodes of vehicle dynamics are basically how you learn to drive your MR2, so rather than R-compound tires, for the novice autocrosser a decent set of performance street tires are better for learning. Photo courtesy Bryan Heitkotter.

Your local autocross venue should offer a pit area, basically a place where you perform last-minute, on-site preparations. Photo courtesy Bryan Heitkotter.

This is especially critical if you're going to drive home on the tires you race with. With the ever more popular street tire racing classes, combined with the progression of tire technology, more and more enthusiasts are driving on the street and at the event on the same set of tires. Tire blowouts, which are much different from a simple flat, are extremely dangerous!

On-Site Event Preparation

When you arrive at the autocross course:
- Verify registration
- Empty car of everything not bolted down, including spare and jack, turn off radio
- Swap wheels/tires, if necessary
- Check air pressure in tires, torque on wheel lugs, and chalk tires
- Pass tech inspection
- Be sure your car has visible number/class given to you at registration or tech inspection
- Adjust shocks
- Turn mirrors off
- Adjust seating position to appropriately address steering wheel and pedals
- Warm engine to proper operating temperature before first run

There are generally three levels of autocrosser: the novice that just shows up—run what ya' brung, so to speak; the more serious, prepared autocrosser who runs on R-compounds, chalks his tires and makes other adjustments but is self-contained in his own car with maybe a tire-cart trailing behind him; and then there are the people with a lot more money that trailer their cars—anything from very streetable Miatas and RX-7s to extensively built Dodge Vipers and Formula Fords. (Those people are cool.) The only thing that is absolutely required to be successful, though, is a stick-with-it, figure-it-out, have-fun attitude.

When you arrive at your autocross, usually a weekend during the spring, summer or fall, the first thing you should do is your on-site prep. This preparation is in addition to what you've done in the weeks leading up to the event and is limited in scope. It basically amounts to emptying your car of anything that isn't bolted down, including floor mats, CDs and that crown-shaped air freshener on the dash. If you have adjustable shocks—and you do, don't you?—make sure they are set where you like them, and check your tire pressure and chalk your tires as well.

This SW20 has "98ASP" written on the rear quarter-glass. "98" is the driver number, while "ASP" reflects the car's "A" classification, and "Street Prepared" modifications. If the car were stock, it would simply have read "98AS," for "98 A Stock." Dating the picture a bit, the SW20 has since been moved to class "B" by the SCCA.

171

Toyota MR2 Performance

Officially identifying your car requires a number, usually followed by a class designation (missing in this picture). You can use shoe polish on windows, or tape or magnets on the doors—as long as it is legible from a distance, it should pass tech inspection. Photo courtesy Bryan Heitkotter.

Note: If you're changing out your tires at an event, be sure to have a torque wrench; "pretty darn tight" isn't precise enough. All generations of MR2 should have their lug nuts torqued to 76 ft-lbs. On the SW2x, being 5-lug, be sure to apply the torque in a "star pattern," for example, starting at the bottom left, going to the top right, going to the left middle, and so on, always going as far away from the previously torqued lug on each following lug as possible.

Turn The Mirrors Off—This one can be a little odd at first, but there is a purpose here. If you're on an autocross course, turn the mirrors off. Twist the rearview mirror up and out of sight, and fold in your side-view mirrors. Cones leaping up into the air as you tear past is not the confidence boost you'd think it'd be. Your focus needs to be on the next turn, or event, not on what's behind you.

Unless of course you're on a road course. In that case, you'll need your mirrors to watch for faster cars needing to pass, and drivers in slower cars celebrating your speed (in their own way) as you pass. The idea, though, in either case is to try and keep your eyes forward, always looking out and ahead.

SCCA Classification Rules and Regulations

In an attempt to level the playing field, cars are grouped into classes by the SCCA. Classes are labeled using letters working from A down, so that an "A" class car is expected to be faster than a "B" class car, and so on. Power-to-weight ratio, balance, tire type and size, and a slew of other factors all play into even a completely unmodified car's performance.

The SCCA is not shy at all about reclassing cars as they see fit. Many autocrossers take this competition very seriously as well, including selling one car and buying another if a reclassification renders a car non-competitive in its new class. Personally, I have always approached autocross as an opportunity to find my limits as a driver, and my MR2's limits as a sports car, so that I build my MR2 as I want it, and run it wherever it is classed regardless of its "competitiveness" in that class—more on that later.

The following information is a verbatim excerpt taken directly from www.scca.org regarding stock class rules and regulations to give you an idea of what the SCCA considers "stock," and a quick look at some of the types of modifications allowed across classes. For the most current exact language and listing, see the scca.org website.

Stock—This category includes mass-produced, common vehicles that may be "daily drivers" (cars used for normal, everyday driving). Stock category cars compete in their "factory" configuration with a minimal number of allowances (not requirements) such as:
- Removal of spare tire and tools
- Front anti-roll bar(s)
- Suspension/wheel alignment using standard adjustments
- High-performance DOT tires (including competition R-compounds)
- Shock absorbers/struts (2 external adjustments maximum)
- Competition-type seat belts (no shoulder belts in open cars)
- Brake linings (pads/shoes)
- Air filter element (the "throwaway" part)
- "Cat"-back exhaust systems
- Wheels of standard size (diameter, width, and offset within ¼")
- Roll bar/cage
- Gauges, indicator lights, etc.

Street Prepared—The original SCCA "street" category, Street Prepared allows any carburetor/fuel injection system and any ignition system. Turbo/supercharger hardware has to remain standard, but aftermarket boost control systems/programs and intercoolers are allowed. Exhaust manifolds and systems are free. Emission controls are not required for competition, but no internal engine or transmission modifications are allowed beyond factory specs (no cams, hi-comp pistons, ported heads, etc). Some cars are able to update/backdate components like engines, brakes, etc (See Solo Rules for details and specifics). There are no limits on wheels sizes or DOT tires; racing

Koni shocks are "infinitely" adjustable by turns of a knob. This also means you must keep track of the turns though, as there is no visual indicating current stiffness. Photo courtesy Bryan Heitkotter.

Tokico Illuminas are 5-way adjustable, adjusting both compression and rebound at the same time. A very skinny flathead screwdriver is necessary to make adjustments.

springs and shocks are the norm. (Information sourced from www.scca.org.)

Basic Tire And Suspension Adjustments

For pure performance, our staged guide and a unique "autocross guide" would not be much different. Performance is performance for the most part, with small changes—brake fade being less of an issue autocrossing than other high-performance venues for example, while a massive front sway bar on an SW2x would probably serve you better autocrossing than the water-injection system recommended in later SW2x stages.

The most significant difference would be that in autocross, a higher premium should be placed on consistency, adjustability and grip. Horsepower is far less important than repeatable performance, higher levels of grip and agility, and the ability of your MR2 to rotate in a driver-induced, controlled manner. And beyond these differences, there lies a big pile of rules and regulations that you absolutely must master if you want to be competitive—in the autocross world, everything depends on what class you're in.

The modifications you are allowed in each class are strictly dictated by the language of the SCCA regulations, so a truly helpful set-up guide could stretch to fill an entire book of its own, all based very closely on the available SCCA rules and regulations, a document whose 2008 version is 336 pages long.

Suspension Settings Starting Points—In regard to between-runs suspension setting adjustments, you're mainly limited to tire pressure and shock settings. And these are the only suspension adjustments available to the stock-class MR2 owner, so we'll focus our attention there.

Autocrossing an AW11, fairly neutral suspension settings front to rear can be used, though, again, this is highly dependent on your driving style. Anytime you change anything on the car, whether by adjustment or outright replacement, as obvious as it may sound you should be attempting to achieve some specific end. Before you bolt on anything new to your car, ask yourself what behavior you are trying to change.

To do this requires that you have some impression of what the car is doing before that change or adjustment. What is the car not doing that you want it to do? Since AW11s are keen on understeering, even more so in the later, revised 1987–1989 version, you'll want a more aggressive alignment than you might have on a street-only MR2. Early AW11s seem to benefit from more even shock settings all around, and later AW11s liking stiffer settings in the rear and softer up front.

With the SW2x, stiffer in the rear, softer in the front, depending on driving style, and other settings, like tire pressure, spring rate, toe and camber settings, and sway bar stiffness. To establish the ideal setup for your SW2x, set your shocks nearly full stiff—or full stiff—all around, and then soften up the front between runs to get the results you're looking for, being sure your tire pressure is close to even all around, and that your alignment is close to specification. For Tokico Illuminas, this would mean setting them at 4 or 5 front and rear, and adjusting from there.

You're after a suspension that is predictable in its understeer and oversteer, a car that sets quickly in sweepers, but is not so stiff that the communication from the tire on the pavement to the palms of your hands is muted.

Same with tire pressure: as we mention, setting pressure even all-around at a high pressure, 40 psi

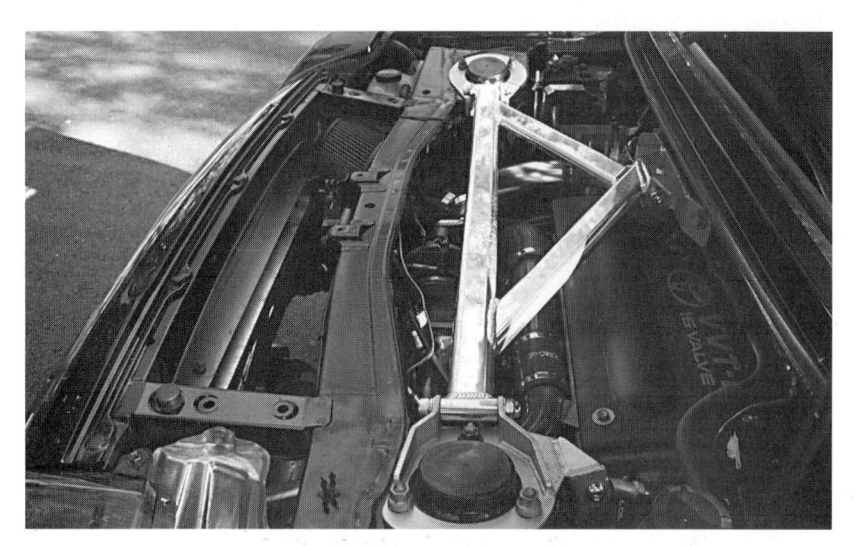

Enhancements to the structural rigidity of your ZZW30, like this strut tower bar, will not only enhance ride comfort will really shine in situations where the chassis is really loaded up, as it is on an autocross. Photo courtesy Dave Kerr.

for example, then using the "data" from your tire-chalking and your backside to allow you to make small tire pressure adjustments between runs.

For the ZZW30, initially softer balanced settings front and rear are a good starting point, on Koni sports running them from quarter of a turn from full soft to full soft, and adjusting from there based on the behavior of the car on course, with tire pressure adjustments similar to the AW11 and SW2x.

Tire Pressure—Your racing tire pressure should generally be higher than your daily driving tire pressure. Increased tire pressure basically strengthens and supports the carcass, keeping your tire from rolling over on its sidewalls. Increasing tire pressure also effectively increases the functional spring rate of your suspension as well, underscoring how much room there is for improvement in these sorts of optimizing, free modifications.

Though as a general rule you will also increase the responsiveness of the tire as you add air, the best pressure for your tires depends on the way your MR2 is handling, the way the suspension is set up, and the type of tire you have. This is not a case of "higher is always better." There is an ideal pressure where your tire will provide optimal grip at a certain temperature, on a certain surface type. By adding or removing air, you are actually trying to reduce the grip in very small amounts in a controlled, intentional manner. In this way, you can induce or reduce oversteer as necessary. Handy, huh?

Though the lion's share of grip is provided by the broad footprint of the tire, much of tire performance boils down to the innocuous tire sidewall. Performance tire manufacturers spend a lot of time trying to figure out ways to enhance the strength of your tire's sidewall, Goodyear, for example, using carbon fiber to do so. Less energy is wasted by the tire deforming under load, providing improved steering response over an underinflated or soft-sidewall tire.

The best way to establish the ideal air pressure for your MR2 boils down to a simple two-step process. Start with a higher, consistent pressure front to rear, say 40 psi, and removing air pressure accordingly 2–3 psi at a time between runs to fine-tune the handling bias. If you have a R-compound tire that you have experience running at higher pressure, starting within 6–12 psi your known sweet spot, and using chalk and oversteer/understeer data to help you further fine-tune things is acceptable as well. Taking some pressure out of the rear tires and/or adding some to the front can help reduce understeer, and the opposite can render some additional oversteer—whether you use higher or lower pressure to manipulate grip levels is up to you.

Note: A fully stock MR2, especially older generations with some miles, will probably not have enough negative camber to maintain optimal contact with the road during aggressive cornering procedures. Inevitably, no matter how much air pressure you add to the tires, some MR2s will roll over on their sidewalls anyway. At some point, what you need is improved roll control, not more air in your tires.

Note: Lighter MR2s like the AW15 and ZZW30 may require less air pressure than heavier MR2s like the AW16 and SW20.

While the softness or stickiness of the tire compound gets all the attention, tire pressure and sidewall contribute heavily to the on-course performance of your MR2. Pictured are tires with sidewalls much shorter than the factory tires, increasing the sharpness turn-in, possibly at the expense of breakaway warning. Photo courtesy Hsun Chen.

AUTOCROSSING YOUR MR2

This Bridgestone Potenza all-weather street tire is never going to be able to stay off of its sidewall like a stiff-sidewall R-compound tire. You can just make out the scuff marks almost all the way down to the bottom of the "B" in Bridgestone. This tire is all over its sidewall under aggressive cornering.

Chalking your tires will help to illustrate what the tires are doing under hard cornering—the sidewalls obviously have significantly less area and grip than the shoulder of the tire itself. A half-dollar size mark at several points around the tire will give you the information you need. Photo courtesy Jensen Lum.

Note: Radical tuning should not be done with tire pressure—nor with spring rates, shock settings, alignment, etc. Remember, we're fine-tuning here.

Note: Tire pressures increase 1 psi for every 10 degree increase in outside air temperature.

Chalking Your Tires—To help figure out the ideal tire pressure for you MR2's tires, you need a mix of data, and good old-fashioned feel. The data comes from chalk, or shoe polish. In order to find out how well the tires are holding their shape and staying off their sidewalls under aggressive cornering, you can mark the tires on the sidewalls, and see how if and how it scuffs under cornering.

To chalk your tires, take some white shoe polish and dab the sidewall on the top of the tire, right near the edge of your tread. This will give you a rough idea of what the sidewall is doing in turns—if most of the shoe polish is gone after a run, you need to add some air pressure. If none of the polish is gone at all, too little pressure isn't an a problem. At this point, you can begin removing 3–4 psi between runs until you see the polish scuffing some—then you can go back up, and sit tight.

Chalking tires is a quick and dirty method of analyzing the way a tire's serving your handling needs. It could get a little more complicated too if you're interested, involving measuring tire temperatures with a tire pyrometer.

Tire Pyrometer—A tire pyrometer, which measures the surface temperature of the tires, is a helpful tool to assist you in fine-tuning your MR2's handling setup. The idea with any suspension setup is to make sure you have as much tire patch contacting the ground as possible. This contact, married to a few factors, namely tire pressure, weight, compound, and wheel alignment, creates friction, and friction creates heat.

Note: A pyrometer is a wonderful tool to find out a lot about your tire, alignment, suspension setup, and even your own driving style. A pyrometer featuring a memory that recalls recent readings so you don't have to is highly recommended.

A pyrometer allows you to take the temperature of the tire after a run to basically infer what the tire is doing while you're busy driving—how much stress it is seeing, and where, can tell you a lot about the handling balance of your MR2, and your driving style as well. A few tire-temperature tips:

• Front readings may be slightly higher due to "artificial" heating from the front brakes, while on the MR2, this may be balanced out by its mid-engine design that may heat the rear tires.
• A reading should be taken immediately after the car stops, on the inside, middle, and outside of the tire. For example, a reading of 180°F across the tread would be ideal, indicating an even friction—and surface contact, rather than temperatures increasing and decreasing in an anomalous manner.
• Higher front tire temperatures can indicate excessive understeer.
• Higher rear tire temperatures can indicate excessive oversteer.

TOYOTA MR2 PERFORMANCE

Part	ADJUSTING HANDLING BALANCE Reduce Understeer	Reduce Oversteer
Tires		
Front Tire Pressure	Increase	Decrease
Rear Tire Pressure	Decrease	Increase
Shocks		
Front Shocks	Softer	Stiffer
Rear Shocks	Stiffer	Softer
Springs		
Front Spring Rate	Softer	Stiffer
Rear Spring Rate	Stiffer	Softer
Sway Bars		
Front Sway Bar	Softer	Stiffer
Rear Sway Bar	Stiffer	Softer
Alignment		
Front Camber	More Negative	More Positive
Rear Camber	More Positive	More Negative

After the preparation is complete, it's time to play with your MR2. This SW21 patiently waits its turn, while drivers in the background walk the course. Walking the course is done to familiarize yourself with turns and other events you'll encounter. Some drivers skip this, while others literally count their steps. Find what works for you, but be sure to try it both ways. Photo courtesy Bryan Heitkotter.

• A cooler middle can be a sign of too little air pressure. If the outside of the tire is cooler than the inside, it indicates that the tire is seeing a less-than-ideal angle with the road, so you should be checking your alignment settings, especially camber.
• Tires cool rather quickly, so when you're using your pyrometer, take the reading as soon as possible after the run.
• You also can find an empty parking lot and simply run in circles so that you can gain "pure-turn" data—the same concept as a skidpad. In the case of an autocross for example, the temperatures you are seeing are also an average of both straightaway and corners, so take that into consideration if you are gaining readings at the event.

Driver Input

At this point, your MR2 has been prepared well in the weeks leading up to the event. You've arrived early enough to walk the course if that is part of your routine, and do any necessary on-site preparation. Now it's time to race.

The Steering Wheel—How you hold the steering wheel depends first on the seat's position. Having your seat too far back requires exaggerated arm movement, and is more fatiguing. Having the arms bent gives you more leverage and allows you to turn the wheel with smaller, smoother movements. Address the steering wheel so that your hands can grab the wheel at both points while still bent and relaxed. Your feet should also be able to manipulate the pedals easily without extending fully.

Once positioned well, control the wheel at 9 and 3 o'clock. Do not squeeze the wheel—a lighter, consistent pressure is ideal. When you're driving, your hands should only leave the steering wheel to shift, and then just long enough to shift. Your hands should rest at the 9 and 3 o'clock position, though some drivers prefer an almost 10 and 2 position. This hand positioning goes for turns, too. As you approach a turn, a 180-degree "U" to the right for example, you should feed the wheel from your left hand (more or less) at 9 o'clock, to your right hand (more or less) at 3, with both hands on the wheel throughout. As you find your apex, point the car and begin accelerating out of the turn, straightening the front wheels in the process, your hands should pretty much stay at that 9 and 3 position. It is the wheel that moves.

Of course they will move some, but you are trying to maintain consistent angles with your hands so you can familiarize yourself with the relationship between the geometry of your hand-positioning, and the reaction of the car. This is especially helpful if you are trying to regain control of the car, and everything is happening in a hurry; having a natural, instinctive "home base" hand positioning is extremely helpful. This keeps you from thinking too much, and allows you to react with trained instincts that are quicker than calculated thought.

How you hold the steering wheel is important for maximum car control and precision. The 9 and 3 o'clock is the classic, tried and true approach. This grip allows for steering input free from sawing, and controlled feeding of the wheel from hand to hand in tighter turns. When you're driving, your hands should only leave the steering wheel to shift, and then just long enough to shift.

This Miata is understeering in this left-hand turn. Notice the suspension weighted heavily to the outside wheels, and somewhat to the front. When the car is understeering, the front wheels lack enough grip to deliver the lateral motion you are asking for through the steering wheel. Photo courtesy Bryan Heitkotter.

It also helps with smoothness. While it is tempting to simply release your grip and let the wheel fly through your hands back to straight-up, the smoothest movements will have this process going on incrementally through the turn. There are other techniques, of course, and pros and cons to each. Talk to other drivers, do some homework, and try out several techniques to see which works best for you. Just make sure you're using both hands.

Left-Foot Braking—The value of left-foot braking depends on driving style and preference. Many drivers swear by it, while others can whip out a calculator and show you the math of why it is a bad idea. One benefit, driving an SW20 for example, is that left-foot braking will allow you to maintain boost throughout the turn for improved corner exit.

Another benefit of left-foot braking, and probably the more widely applicable and useful advantage, is the ability to minimize front-to-rear weight transfer. Smooth driver inputs are key any time you are driving with pace: the brake pedal should be squeezed on, as should the throttle. When coming off the throttle, this should also be done smoothly, and turns should be made as gradually as the course will allow. The less unsettled the car is, the faster you will be. Applying brake pressure with your left foot, and throttle with your right, and coordinating efforts between them you can help maintain not only the appropriate rate of acceleration or deceleration, but also producing less weight transfer and smoother movements as well.

Being "Early," Looking Ahead—We will get some autocross advice from former national champions later in the chapter. For now, we'll end with two basic autocross rules: Be early, and look ahead.

Looking ahead on the course is just as critical as grip management—in fact, it precedes it. As you exit one corner, you should already be looking towards the next turn, Chicago-box, or slalom. If you've done so, the car will have a tendency to gravitate towards it in a smooth, natural way. This is not possible if you are constantly surprised by the cones. Always be early—if you're "late" for the first cone in a 4-cone slalom, the slalom has been lost. Always be early.

Also, try to be as fast as possible on your first run! I've thrown away the first two or three runs for years because I just "knew" I'd be slow during those runs anyway. I took that for granted, because 99 times out of 100, my last run was my fastest. This may have been a self-fulfilling prophecy, though. Consistency is key to successful autocrossing.

Grip Management

Understeer and Oversteer—Understeer basically means that the car is changing direction less than the steering wheel angle would seem to demand, under-turning if it is turning at all. The front end of the vehicle tends to break loose and slide toward the outside of the turn, often referred to as a push.

Oversteer is the opposite, where the slip angle of the rear exceeds the slip angle of the front tires and

Drifting is basically prolonged, controlled oversteer, expertly demonstrated here by Bryan Heitkotter. No matter how cool it looks, though, it is generally not the fastest way around a course. Photo courtesy Sarah Mays.

During an aggressive launch, the front end lifts and the weight transfers to the rear wheels, or "squats." This is a good thing on a rear-drive car, as it places great weight on the drive wheels. Photo courtesy Jeff Fazio.

During braking, the weight of the car "dives" forward, unweighting the rear wheels. This is why it is a good idea to brake and turn separately, especially in a mid-engine car. A light rear end can come around in a hurry during aggressive cornering. Photo courtesy Bryan Heitkotter.

the back end slides outward, as if it wants to swap ends. It is often referred to as "loose." Neither condition is desirable, especially excessive understeer. Many drivers adopt a style with a slight bit of oversteer that can be controlled by modulating the throttle to help the car rotate the car as necessary, but this is an advanced skill that requires a lot of tuning and practice.

A quick crash course in grip basics: Your four tires provide the four points of contact your MR2 has with the road. Each contact patch has a certain amount of grip to offer, depending on the friction between it and the traveling surface.

That grip can be applied in four directions, and in various combinations: braking, accelerating, and turning left and right. Think of grip in terms of percentages. If a tire's total grip potential is 100%, once you pass that—ask for that patch to do something that overwhelms its available grip—it is going to stop doing something, perhaps dramatically: spin wildly under acceleration, under or oversteer, or lock-up under heavy braking. The key to high-performance driving is consistently approaching and using that 100% (of available grip), only exceeding it when it is advantageous (trail-braking, for example). Leaving grip on the table, or over-driving the car is simply wasting time.

As you accelerate, the car "squats," and weight transfers to the rear. This is one of the major problems of front-wheel drive, as unweighting the drive-wheels causes excessive wheel-spin. On a rear-drive car, though, such squatting causes the rear tires to "dig in," and to a point, increases their grip. Since they are responsible for accelerating the vehicle, the front tires are free simply to turn, which is what makes rear-drive cars feel so well-balanced. This effect is magnified on MR2s, with the weight of the engine placed over the drive wheels.

In contrast to weight transfer rearward under acceleration, braking transfers the weight forward. This dissolves grip over the rear tires, which is especially hurtful if you are at the limit of your tire's adhesion in a turn (but is the concept behind trail-braking). The common scenario is to enter a turn too fast and, upon sensing your mistake, you hit the brakes. This transfers the weight forward, unloading the rear (grip) and causing the car to spin. You only have so much grip to do what you want, and any load transfer brought about by turning left, right, accelerating or braking, uses some of this grip.

Braking: weight transfer forward (dive)
Accelerating: weight transfer rearward (squat)
Note: Trail braking forces oversteer through the unweighting of the rear tires when the front wheels have designated turning angle during braking.

Most forms of automotive competition boil down to effectively managing the grip of your four available contact patches. Understanding this concept is absolutely critical to becoming a fast, consistent driver.

Driver Action, Grip Reaction—At all times on an autocross course—I have tried and cannot think of an exception—you should either be accelerating or braking, one or the other. Steady-state sweepers and similar course paths might require a somewhat steady-state application of throttle, but maintaining constant speed through a turn is technically acceleration anyway. The principle remains—you should always strive to be accelerating or braking, one or the other. Resist the bad habits of coasting, engine braking, etc.

Picture yourself in your MR2 on an autocross or road course, at the end of a straightaway coming upon a turn. You are carrying too much to navigate the turn, so you squeeze the brakes. If you are braking at the absolute limit's of your tire's adhesion—and it is typically the grip of your tires that limits your braking—you are using at or approaching 100% of the available grip of the front two contact patches, which leaves none for turning. Not good.

Ignorant of this fact, let's say you turn the steering wheel anyway. The poor front tires, already overworked, cannot do what you're asking. They are expected to provide grip in transforming the steering angle into lateral motion, but their potential—their grip level—is at zero; at that moment, under those circumstances, they are spent. What happens next is predictable, and common: the car understeers.

Disregarding the option of trail braking, which is basically braking with the wheels somewhat turned to encourage the car to rotate, it is safe to assume that the car will simply plow straight ahead, turning very little. Luckily, this is wasting energy, and you are decelerating, which is exactly what you need to do at that point. By trying to both aggressively turn and brake at the same time, you are going to do both poorly.

Once the car slows enough for the tires to provide enough grip to navigate the turn, the car will begin to turn. Yay! However, with the newfound grip, you get a false sense of security and power, and begin to accelerate through the turn too early. Too much throttle, too soon. This increases the lateral load on the outside tires.

Remember, think in terms of percentages. If you are at your MR2's handling limit while turning without throttle, 100% of available grip is spent attempting to transforming steering wheel angle

Notice how this ZZW30 driver is looking ahead, where the turn is going, rather than directly out in front of the car. Photo courtesy Hsun Chen.

This AW11 appears to be accelerating and turning at the same time, evident by the slightly rear, and slightly driver-side loading on the suspension. Accelerating out of turns is something an MR2 of any generation can do exceptionally well due to its mid-engine, rear-drive configuration—just don't lift mid-corner. Photo courtesy Bryan Heitkotter.

into lateral motion. As you add throttle, a percentage of that grip being used for lateral grip is basically stolen by acceleration. This is only made worse considering that the added speed provided by the throttle is increasing the need for lateral grip.

This is the basic conundrum of weight transfer and grip management. Trying to do too many things at the same time, trying to do too much too soon, and not being smooth in how you are "doing" any of it. The earlier you can get your braking done and the apex clipped, the quicker you can get to accelerating at the car's full potential, effectively extending the straightaways.

You have to be several steps ahead of the course, not behind. Prioritizing the inputs and subsequent

Snap oversteer is different from old-fashioned, tail-out oversteer, and avoiding it amounts to not overcorrecting as the rear initially steps out. Smooth is fast. A slalom is an easy to place to "create" a snap oversteer situation. Mastering the slalom is a matter of staying early at each cone, and establishing a smooth rhythm with minimal steering inputs as the slalom progresses. Photo courtesy Hsun Chen.

dynamic movements of your car is key to fully realizing your MR2's potential. A fast driver will use all of the tire's grip potential in a consistent, intentional manner.

A Final Note on Autocrossing

Personally, I view autocrossing as an opportunity to compete against the course, my own abilities, and ultimately the potential of my car more so than a competition against other drivers. To be clear, this is only one approach, but it an option.

Have fun autocrossing—it's a blast. And if you mean to compete successfully on a local or national level, enjoy that. Just keep in mind that simply competing against yourself—and the abilities of your own car—is also a worthy endeavor. Building your MR2 exactly how you want it and extracting maximum performance, no matter the class, can be a perfectly satisfying approach to all forms of amateur motorsports. Don't let ego and competition cloud the real fun of it all, as it can be so easy to do.

In a slalom, once you've ruined your timing and rhythm, it's difficult to regain it while staying fast and missing cones. Hitting a cone, especially during your first or second run, is not the end of the world, showing that you are using all of the physical space alloted you by the course designer—and then some. Finding the limits often requires going just beyond them, as this SW21 driver found, hitting two cones out of the six-cone slalom. Photo courtesy Hsun Chen.

Internet Driving School Resources—On the Internet, beyond your favorite search engine or automotive community of fellow hotshoes, the following websites are good places to start when looking into honing yourself into a high-performance driving master of the universe.
www.scca.org
www.nasaproracing.com
www.skipbarber.com
www.bondurant.com
www.jimrussellusa.com

Autocross is simply a professionally organized opportunity to go fast. People autocross for a variety of reasons: competition, camaraderie or simply the exploring the limits of a sports car. This would apply to road course and drag racing as well. Photo courtesy Bryan Heitkotter.

A CONVERSATION WITH TWO MR2 AUTOCROSSING PROS

Note: Randy Chase and Randy Noll are two drivers who have competed MR2s successfully on a national level. Each driver has their own style, level of experience and approach to car setup. The following tips have been used by both of them to successfully campaign MR2s on a national level.

Randy Noll's Tips

2001 Solo II National Rookie of the Year; 2nd ES 2002 Solo II National Championships; 2002 ES ProSolo National Champion; 3rd BS 2005 ProSolo National Series; Numerous National and ProSolo wins; Driving Instructor

1. Something I call "on-course consciousness." It's the opposite of the "red mist" they say overcomes a driver once he's on course. It's being able to think, react, absorb, adjust and learn while you're driving, and remember what happened on course when you're done. It's a huge hurdle to overcome but is key to learning to perform in any sport that happens very quickly. As hard as it is to hold back, the easiest way to achieve on-course consciousness is to slow down until you do it, then gradually push harder while continuing to actively think while driving.

I think most top drivers have long lost the thrill of speed or sliding. They are driven by the thrill of execution. It's the designing of and building of the roller coaster—the ride is just a by-product. A truly great lap is awesome in the fact that it is uneventful, almost boring.

2. The most important corners are the ones with the longest "straights" after them. A "straight" is a piece of track where your foot is all the way to the floor. If you can nail your apexes and throttle spots on the five corners before the five longest straights on course, you will go fast.

3. Position over power or aggression. More than anything, hit your apexes. Not just close, nail them. If you hit your apexes, you're probably doing many other things correctly.

4. Early throttle spot. Corner exit is 100 times more important than corner entrance. If you get on the gas five feet before point A, you'll be going a few mph faster when you reach point A, and a few mph faster for the rest of the straight after point A.

5. Take time right after the run to evaluate every portion of it that you can remember. A codriver can be a huge help with this by asking you questions about the run when you get in. Learn from every run. A quick rule of thumb: If a portion of the course felt difficult or cramped, you probably didn't slow down enough for it. If a portion of the course felt too easy, you could probably push harder/faster there next time.

6. Find a good instructor. A fast driver is not always a good instructor. When you find one, let go of what you think you know and take all of their suggestions with an open and unbiased mind. Driving fast is counterintuitive for most people; it must be learned. Embracing and applying new instruction is the fastest way to get rid of bad habits and begin building good ones. Evolution Performance Driving School is the best school I've encountered to date.

Randy Chase

2000 CS ProSolo National Champion; Numerous ProSolo wins; Driving Instructor

I have heard and learned many important tips over the years. Sometimes I wonder if you have to be ready for them...as if maybe I heard them before, but I was not ready. What follows are various tips that made a significant difference in my driving at the time I heard them:

1. As soon as you pass the start line, you are only losing time. Minimize how much time you lose. It's not just going faster, it's spending less time on the course.

2. The trick is to drive 10/10ths. A novice will drive 7/10ths and then not realize that they went to 13/10ths. Learn the edge and drive it. That may mean some cones get hit and you spin. It's what teaches you where 10/10ths is.

3. There are fast parts and there are slow parts. Learn the difference.

4. Don't square off the corners and point and shoot drive. A lot of corners are parts of smooth arcs you can make. It's faster to drive a smooth arc than a short straight and two jerky turns.

5. Be aggressive in chicanes. Attack them, stay in front of the turns and as straight as you can. Getting "behind" in a chicane is a bad thing.

6. Doing this well means being smooth. Being smooth does not mean you are slow. To drive the car smoothly may require controlled chaos in the car. Fast hand and foot movements do not mean you are not smooth.

(continued on next page)

(continued from previous page)

7. Know your line you intend on driving. Understand it. Look for it. If you drive such that you are forcing yourself off that line, you made a mistake and need to slow down. The line is everything, unless you are wrong about where the line is...then you need to change your mind.

8. Look ahead to where you want to exit the turn. Adjust your speed into the corner to make sure your car will be on the right spot when your exit the corner.

9. It's better to corner under acceleration than braking. Brake earlier and then get on the throttle as quick as you can.

10. The earlier throttle points will be faster. Give up the end of the straight to make your corner exit faster. That speed coming out of the corner will carry through the whole straight following the corner.

11. Do not try to save runs once they go bad. If you get screwed up, relax. This saves tires. Focus on driving the rest of the course correctly. You can't fix that run by going faster on the rest of the course.

12. Understeer is often caused by going in to a corner too hot or too much front tire slip angle. To reduce understeer, straighten out the steering and/or reduce throttle input (or often, just brake earlier).

13. A lift or quick tap on the brakes can cause the front of the car to weight and allow better turn in.

14. If you start going slower or are less successful than you should be, check the car. Sometimes things change and it's hard to notice.

15. Don't worry about memorizing long lists of tips. Work on one or two things at a time. Don't try to adjust everything, put in a new sway bar and struts, try out Hoosiers, and decide to use left foot braking all in one weekend. Make changes one at a time and see how they feel. Spend more time on your driving than your car setup.

Chapter 8
An MR2 Buyer's Guide

Choosing the right MR2 requires research, patience and honest reflection about what you want from your sports car. This beautiful 1994 SW20 represents an interesting find due to the rarity of its Tropical Blue color. Photo courtesy Osman Ullah.

USED MR2 AVAILABILITY

MR2s are versatile cars, capable of excelling as daily drivers, track toys, and weekend cars for grinning and touring, the way a proper sports car should. It is not necessary for sports cars, as I see them anyway, to be blistering fast, though they certainly may be. Rather than ability, the definition of sports car as I see it has more to do with purpose. With that purpose comes a used MR2 market that offers an extremely wide range of quality and options.

The MR2 market has changed drastically over the years. The SW20 that sticker-priced itself right out of business over a decade ago is now quite affordable, and accessible for folks whose main form of transportation not long ago may have been a skateboard. Decent AW15s can be had for as little as $1500, AW16s and SW21s for $3500, SW20s for $4500, and ZZW30s for under $10,000. While this is a boon for the intending sports car owner of average means, it also creates market issues.

AW11: Rust and Affordability

Rust is probably biggest long-term threat to the AW11. Finding examples without rust is becoming increasingly difficult, and rust, unlike a worn clutch or even a blown head gasket, is not easy to deal with, addressing the issue usually requiring welding in new sheet metal to adequately address.

This corrosion aside, its most significant threat would have to be the threat of affordability. The first-generation is so affordable that many are bought not as MR2s, but as simple

The formerly prohibitively expensive SW20 is now changing hands with folks who could not afford one new. This is both good and bad. Photo courtesy Brian Hill.

transportation. Toyota has a magnificent reputation for building durable, quality cars and deservedly so. Unfortunately, though, would-be owners who might otherwise be scared off by the lack of storage space and mid-engine are instead attracted by the affordable price and durable reputation. Often, these aren't owners that are interested in the meticulous maintenance a sports car of this vintage requires to maintain its condition.

There have been recent examples of museum-quality AW11s popping up on eBay and in MR2-specific trading forums. The asking price of many of these is pretty high—

Toyota MR2 Performance

This 1985 MR2 has escaped rust by spending most of its life in California. Rust, rather than mileage, should be the top priority when shopping for AW11s. Trunk space is surprisingly accommodating on both the AW11 and SW2x.

The market for quality AW11s is steadily decreasing. It has been two decades since the last AW11 was sold new in America. Photo courtesy Bryan Heitkotter.

The last AW11 rolled off the assembly line nearly two decades ago, which makes finding a clean example a challenge, more so with every year that passes. Photo courtesy Bryan Heitkotter.

over $12,000 for a 30,000-mile supercharged 1989 model, for example. The more critical point, though, is that they are indeed out there—these vintage MR2s, which means that some owners are indeed collecting and storing them, good news for the MR2 market in general.

SW2x

One of the SW2x's most alluring features is undoubtedly its body style. Regardless of mechanicals or performance, the SW2x can manage to get by on its looks alone even today, its bold look born from classic, wind-shaved lines that have aged well over the years.

The SW20's demise was caused by rising sticker prices, which put the car out of reach for most would-be, hands-on owners. As the car aged, though, prices fell, and suddenly more SW20s were making their ways into the hands of owners wanting their chance to play. All of these factors seemed to coincide between 1998 and 2003, causing a relative peak in the SW20s popularity during that time. What does this have to do with buying an MR2 today? During that window, the stock of available SW20s was affected by owners who may have been a little ambitious and didn't really know what they were doing.

The MR2 was never produced in Ford Mustang-like numbers to begin with, making every poorly modified, wrecked or otherwise junked MR2 all the more depressing.

ZZW30: A Different Breed

The used ZZW30 market is relatively wide open. These cars are still often in the hands of their original owners, and many are lease turn-ins available for a bargain at major dealer lots across the country. The hardcore aftermarket, while definitely active, still hasn't embraced the ZZW30 like it did the previous generation cars, if for no other reason

An MR2 Buyer's Guide

While rust doesn't seem to be as enamored with SW2x as it is with the AW11, the SW2x's good looks are a more significant issue. Photo courtesy Brian Hill.

Not all MR2s are pampered by car shows and carbon fiber. The SW20 attracted the attention of a lot of owners who didn't properly plan or execute the performance buildups, which took a lot of good used cars off the market. Finding the right MR2 for you might take some time. Photo courtesy Vespremi/Crites/Smith.

The ZZW30 market is a bit different from the AW11 and SW2x market—and not just because they're convertibles. Photo courtesy Aaron Bown.

T-top SW20s are the most common, which is okay because hardcore track junkies—who'd value a hardtop over a T-top—are definitely the minority of MR2 owners. Photo courtesy Brian Hill.

than simple accounting: people are less likely to aggressively modify and race cars they are still paying for.

MR2 BUYER'S GUIDE
1985–1989 MR2 (All)
Potential Problem Areas

1985–1986
- Warped exhaust manifold
- 5th gear pop-out
- Warped front brake rotors

1985–1989
- Rust
- Worn steering rack bushings, ball-joints
- Starter failure (due to proximity to exhaust manifold)
- Broken switches and interior placements
- Missing sunroof shade on cars so equipped (sunroof shades can be hard to find)
- Damaged front lip, especially on 1988–1989 models
- Musty smells from leaking T-top roofs
- Faded red paint

Note: Supercharged models tend to be the most sought-after; they also suffer head-gasket failure more often than normally aspirated models.

To help establish value, verify whether that AW16 you're looking at is a factory supercharged model, or one that has had a supercharged engine swapped in (though the owner will usually tell you, and few examples have received such a thorough conversion that you cannot tell the difference simply referencing the bulleted list above). This should not

Toyota MR2 Performance

Beyond the persistent—and geographically specific—threat of corrosion and rust, you should encounter few significant issues in a well-maintained AW11. Shown in its most potent supercharged form, you would be lucky to find an AW16 this clean. Photo Helder Carreiro.

The AW11 front spoiler is especially prone to damage due to collisions with parking blocks more than anything else.

The extended front spoiler on the late AW11 looks great but is even more prone to damage than the early style. Photo courtesy Bryan Heitkotter.

imply that those MR2s having received swaps are automatically worth less. A 4A-GZE swapped into a hardtop model, for example, can be a plus for many buyers. However, factory supercharged models are generally worth more market-wise. To identify between the two, factory supercharged models will have a VIN beginning JT2AW16, rather than the JT2AW15 of the normally aspirated model.

Note: VIN plates can tell you very useful information, like specific production dates, VIN numbers, intended weight limits, factory tire size, and other production information. Production dates can come in handy when trying to determine if an MR2 is an early or late version of that production year, sometimes indicating that it possesses certain changes made mid-run by Toyota.

1991–1995 MR2 (All)
Potential Problem Areas

1991–1992
- Warped exhaust manifold (turbocharged models). In 1993, a 9-stud exhaust manifold replaced the 7-stud unit as part of the significant revisions of that year.
- Worn transmission synchronizers, mainly 2nd gear. The E513 transmission in the SW20 suffered from weak synchronizers in 1st and 2nd gear, a condition exacerbated by increases in torque, and, ultimately, miles. It was revised with stronger brass synchronizers in 1993 in 2nd and 3rd gear. The S54 from the normally aspirated model presents no real issues in stock form.

1991–1995
- Blown turbo seals (turbocharged models)
- Slipping clutch
- Failing/frozen parking-brake cables
- Poorly matched aftermarket components
- Intermittently failing power steering system
- Power antenna that will not fully extend or retract
- Ball-joints
- Musty smell from leaking T-tops; T-tops in various states of disrepair (eccentric guides)
- Missing sunroof shades on models so equipped
- Worn steering wheels and leather seating surfaces
- Verify all recalls have been performed by calling a Toyota dealer and giving them the VIN
- Faded red paint
- Minor issues like skipping or otherwise stubborn CD players, power antenna that don't extend all the way up or down, are somewhat common, and not

An MR2 Buyer's Guide

Broken switches, especially headlights and windshield wiper switches, plague the AW11—though clearly not this AW11. Note the factory green "SUPERCHARGER" light at the bottom of the tachometer. The LED light above the radio is an aftermarket addition. Photo courtesy Helder Carreiro.

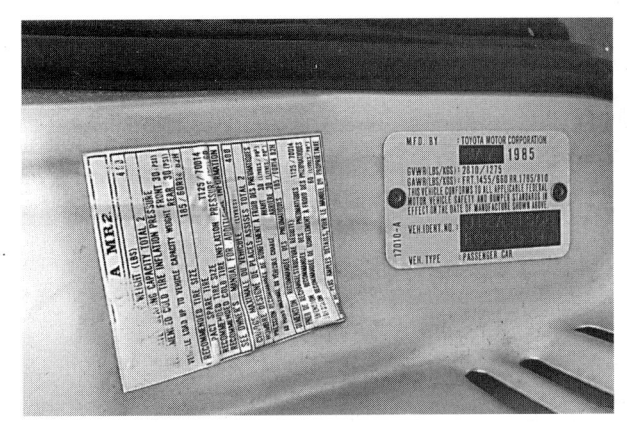

The VIN identifier is visible on this VIN plate from a 1985 MR2—AW15. VIN plates can also tell you very useful information, like specific production dates for use when ordering certain "date-specific" parts.

A factory-correct AW16 will have a trim piece in the front trunk that sits gingerly between the spare and the firewall, another trim piece over the radiator, and a spare tire and jack as well. Photo courtesy Helder Carreiro.

cause for concern beyond their relative cost of repair.

Note: SW20s may produce an odd gurgling sound from the engine bay after spirited driving. Due to the MR2's complex cooling system, with a front-mounted radiator, and twin coolant lines running across the bottom of the car, coolant moves around quite a bit in the MR2, especially the turbocharged model. Coolant may move—or boil—after the engine has been turned off, and this will be easy to hear with no engine noise, and the engine bay a matter of inches from your head.

Note: 1991–1992 cars are much more affordable, being "pre-revision" models. Smaller brakes, a weaker gearbox, smaller, less visually appealing wheels, and the more "lively" rear-suspension bring down the price considerably.

Because of the revisions lacking in the early SW2x, and the rarity of the later, revised 1994–1995 models, the 1993 model year is generally the most commonly sought-after SW2x. Be cautious with the premium you pay for a 1993 model, though. They are improved, but they are not gold.

1993 SW20 hardtops are extremely rare in the USDM. I haven't ever seen official numbers from Toyota, but it is thought that between 25 and 35 1993 SW20 hardtop models were produced, none after 1993.

1994–1994 SW20 are also very rare, and usually command a significant premium.

2000–2005 Toyota MR2 Spyder
Potential Problem Areas
- Pre-cat failure
- Intermittent power steering failure
- Verify the condition of the top

Note: Regardless of year or package, the ZZW30 represents a very similar purchase. If you want a manual transmission, obviously avoid the SMT models; otherwise there isn't a certain year that warrants special consideration assuming you're prepared to navigate the pre-cat issue. The LSD Toyota added in 2004 can always be installed later, and suspension bracing can always be supplemented. As always, simply buy the cleanest example you can afford with the coachwork and options that you are looking for.

Like the AW11, picking the right SW20 may require some legwork, but compared to other mid-engine cars offering similar performance, they present an excellent bargain, and are worth the effort. Photo courtesy Vespremi/Crites/Smith.

This 1993 hardtop SW20 may suffer from excessive wear brought on by years of autocross, but its relative rarity will still command a premium over its far more numerous T-top siblings.

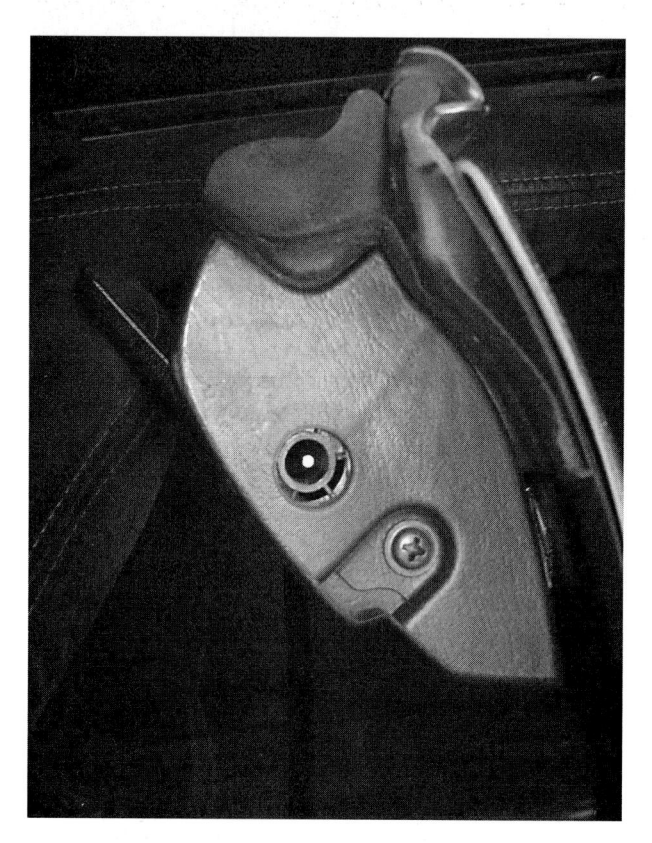

Leaking T-tops are usually caused not by worn rubber seals, but rather T-tops that no longer properly align with the bodywork. This can often be corrected by replacing the T-top eccentric guides. Photo courtesy Jensen Lum.

GENERAL INSPECTION TIPS

An inspection of an MR2 doesn't have to be incredibly complex—most of this is common sense. You are simply trying to roughly determine the condition of the MR2's major components: engine, transmission, suspension, interior, and paint. Secondary components include less expensive and critical components more commonly replaced, like tires, brakes, exhaust system, that should be evaluated as well, but with less concern. With some MR2s, you'll be able to guess from 15 feet away what you're going to find and be correct, while others can fool you.

Engine

Cold Start Behavior—How does the engine act on cold start—one that has not reached operating temperature? Oftentimes a car being warm can hide issues it has. Transmissions shift smoother, valves are quieter, idles are smoother. Always start and drive the car at least once "cold" before purchasing it.

Check for Smoke—Does it blow smoke upon starting, as you apply throttle, or as the engine idles again after some throttling? Blown head gaskets, oil rings past their prime, and bad turbo seals all tend to make smoke.

Check for Unusual Odors—Can you detect the sweet smell of coolant after it runs for a bit? If the owner didn't just change the coolant, it could be a head gasket—and if he has just changed it, you've got questions there too. Changing coolant right before a sale may not be as innocent or helpful as it sounds. Do you smell burning oil? 4A-GE's like to leak at the distributor and the valve cover grommets, which are simple repairs and are not cause for alarm.

AN MR2 BUYER'S GUIDE

POWER STEERING FAILURE

The SW2x had an optional power steering system, an electro-hydraulic power steering system, or EHPS. This system used a separate electric-powered hydraulic pump to deliver steering assistance, which, unlike most power steering systems, ensured that the system did not rob the engine of any of its power. It also offered excellent feel, though it has been known to cause trouble for some owners. If you're power steering in your SW2x stops working—a warning light on the instrument panel should light, and/or the steering effort should go way, way up—there may be an easy fix.

If it is the power steering relay that is stuck, you can try to clear it by turning the ignition from OFF to START (without starting the car—leave the clutch disengaged, etc.) somewhat rapidly, continuously until you've done it about 25 times. This may work. If not, you may need to track the fuse to the power steering motor and check it. It is fused through a fusible link in the front trunk—the ABS 80 amp fuse. If it's blown, you've found your problem. If neither of these fixes do the trick, consult your BGB and have a blast.

1991–1992 cars are much more affordable, being "pre-revision" models, due in large part to the much-publicized handling issues of the early SW2x. Picture is a press photo of the 1991 SW20. Photo courtesy Toyota Motor Sales.

The last ZZW30 was produced for the U.S. market a full twenty years after the AW11 was introduced; it is also far simpler than the SW20, making it and the SW21 generally the most reliable of the MR2 models produced. Photo courtesy Toyota Motor Sales.

The rear main seal in the 3S-GTE can get messy, while 1ZZ-FEs are less problematic in regards to oil leaks. If a 1ZZ-FE is producing smoke, and appears stock, you might suspect pre-cat failure, but obviously try to find where the smoke—and oil—is coming from before you write it off.

Check for Unusual Noises—Go back to the engine, grab the throttle arm and give it a light tug to casually rev the engine. Hold it between 1500 and 3000 rpm and listen for a light rattle within the block—not the clatter of the valves, but something more ominous. Rev it slightly and let it fall again several times to listen for an internal knock, or rattle which could signify worn rod bearings—rod knock. Rod knock is bad.

Worn rod bearings aren't always easy to pick out, but obviously you're going to want to try. And before you assume that what you hear is rod knock, weigh the car's overall condition. If it has 65,000 miles, and the owner has meticulous service records, unless you're an expert at picking out rod knock noises, you can probably assume that what you're hearing is something else, though obviously, you should get it checked out to be sure before spending any money.

Notice the light ticking of a leaky exhaust manifold? Early AW15s and SW20s both like to

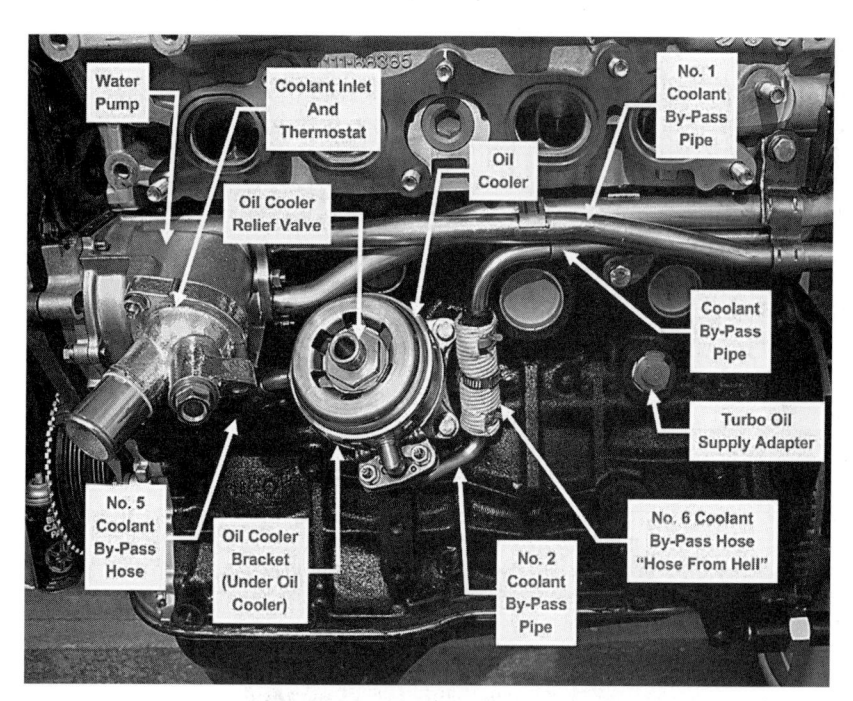

The amount of oil and coolant plumbing on an SW20—with the radiator in front, and a turbocharged engine behind the driver—can make coolant leaks difficult to pinpoint. With the turbocharger and engine sharing oil and coolant, and the relative heat of a turbocharged engine, maintenance on the SW20 is critical.

Pop the engine lid on a stock 1991–1992 SW20, and what you're going to find will look something like this—stock down to dirty fingerprints on the airbox. A well-maintained, stock 3S-GTE should easily last 175,000 miles. Photo courtesy KO Racing.

This 1991 SW20 has had its original passenger door replaced with a red door—and then cheaply painted over. You can just see the red chipping through at the bottom.

warp exhaust manifolds, allowing exhaust gases to leak in a rhythmic "tick." This is not a significant issue in the sense that it is an expensive or particularly troublesome fix. It could be, though, an issue relative to the deal you are brokering specifically. Exhaust manifold leaks are the type of automotive malady best used to get the price down some on a car you're looking at. Don't walk away from a twenty-year old car over an exhaust manifold leak, just consider it as a factor.

The point here is to make careful observations, and do not be afraid of asking questions, or getting a second opinion if you're uneasy.

Test Drive Observations—Drive the car. Is the powerband smooth? Does it return quickly to a smooth idle after closing the throttle? Does it maintain proper operating temperature? Does it emit any odd smells during or after the drive? Unless you've driven several examples of each variation of MR2, it's difficult to know what to expect performance-wise, but you can look for uneven powerbands, jerky acceleration, misses, odd sounds while accelerating briskly, and so on. Many ignition issues can cause problems like these, whether it's an ECU pulling spark timing, or worn plugs and plug wires not able to fully light the air/fuel mixture.

Body

Check for Repainting—If the car has been repainted, what is the quality of the paint job? Is it the original OEM color? Unless it is done exceedingly well, stay away from cars painted with custom colors. Look for drip marks, poorly sanded areas, different colored door-jambs and engine bays, and over-spray and chips in post-factory spray jobs. You may not mind a red engine bay and a silver body, but the person you try to sell it to next might. Reselling is a possibility that we rarely consider while we are in the lust of purchasing.

Check for Body Filler—MR2s get wrecked a lot, mainly because they end up in the hands of poor drivers in over their heads. Be especially diligent

An MR2 Buyer's Guide

It's normal for the middle section of the 3-piece spoilers on the 1991–1993 SW2x to suffer some gap from its driver's side and passenger portions, so this isn't necessarily proof it's been rear-ended by a train. All too often, people use spoilers to close the trunk lid.

The body kit on this SW20 is a replica, rather than the far more expensive (and higher quality) authentic kit. The replica body kit a common phenomenon in the automotive aftermarket where genuine body kits can cost thousands of dollars and can be impossible to find. Note the cracking paint.

checking the rear quarter-panels, as oversteering MR2s often impact there, and they are obviously not replaceable on the AW11 or SW2x (praise Toyota for changing that on the ZZW30, an advancement a long time in coming). You can check for body filler with a magnet—many body shops use pencil-like devices that will stick to factory metal work, but not fillers like Bondo. A car with light repairs isn't the end of the world, but if it has more filler than metal, stay away.

Check Body Panels—Are the body panels OEM? Do they line up properly? Do they still have the factory VIN labels? Are the bolts holding the body panels on new, or do they seem to have been removed/replaced?

Panels that don't line up perfectly don't necessarily mean the car has been hit by a train, but, as with any information you find, you can use it to piece together the big picture.

It's normal for the middle section of the 3-piece spoilers on the 1991–1993 SW2x to suffer some gap from its driver's side and passenger portions. The engine bay lid on the SW2x can also show a slightly irregular gap between itself and the removable panel to its left and right but this is more than likely the result of the car having had these pieces removed several times for work in the engine bay, rather than poor collision repair.

Check for a Body Kit—Aftermarket body kits generally lack the quality of OEM pieces, chipping, splintering and not as sturdy in an accident. They also can fit poorly, and are often improperly installed. If it is a quality kit that has been properly installed—and you like it—no worries.

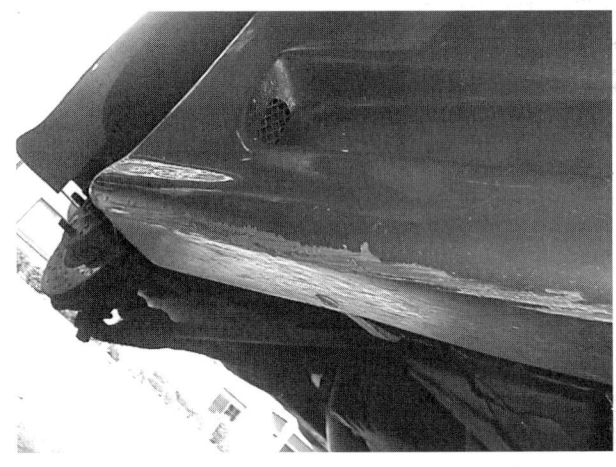

Body kits also generally accept bumps, scrapes, and collision of any sort with far less durability than a factory body piece, while encountering them more often due to a generally lowered ride-height. This TOM's kit got the worst end of a meeting with something that didn't move.

Maintenance History

Check Dipstick—Is the oil honey-colored, or black? If the owner doesn't mind, and you've brought a tool kit, pull a spark plug and check the business end of it as well. A black dust is normal—means you're running rich, which most cars do. The exhaust tip probably has a similar coating on the inside as well. Be sure it is a powdery substance, and not a slime, which indicates oil issues in the combustion chamber. If you're good at reading them, spark plugs can tell you almost everything you need to know about what's going on in the combustion chamber.

Toyota MR2 Performance

MR2 cabins are intimate places, and interiors that aren't up to par can drastically impair your ownership experience. In this clean SW20 interior, note the aftermarket Recaro seats, and removed T-top shades. Photo courtesy Jensen Lum.

Purchasing a modified car can be the soundest financial decision you ever make involving your MR2. It can also turn into an absolute nightmare. The difference maker is due diligence. Photo courtesy Jensen Lum.

Note: Reading spark plugs is a valuable skill, and can tell you a lot about not only the air/fuel mixture, but the timing, and even the condition of the head gasket, valves, and compression rings. They are like cameras in the combustion chamber.

Ask for Service Records—Many owners are through record keepers, and thorough records should be soothing—but not more so than the car in front of you. Receipts are good, but consider more strongly the product you are buying, lest we try to predict the weather without looking out the window.

Check Tires—Tires are like horse's teeth: they can be excellent barometers of not only the car's health, but of its history as well. Owners that keep newer, matching higher-quality tires on their MR2 tend to take care of it elsewhere as well, though they needn't be high-performance.

Mismatch tires (cheap owner), tires with excessively scuffed sidewalls (aggressive-driving owner), and tires worn to the belt on the inside edges (inattentive owner) all at the very least give you something to consider about the recent history of the MR2 you're looking at.

Miscellaneous Checks

Check Interior Condition—Ashtrays can be missing, side bolsters on seats can wear quickly, but a larger concern is missing interior pieces, screws, and other fasteners. If the car has a sunroof or T-top, it should also have removeable sun shades for each, which are often lost.

Leaking T-tops can leave interiors that smell like mildew. The leaking T-tops are usually a matter of T-top eccentric guides needing to be replaced, an easy repair. The mildew is another issue.

Previous Owner History—How long has the current owner owned the car? Again, these are all generalizations that, when weighed together, hopefully will allow you to put together the big picture of the car's history. Cars with fewer owners that have typically been owned longer are often maintained more carefully. Cars that are resold frequently could be a sign of trouble. If the owner has only owned the car a month and is selling it, you've got to wonder why. Many MR2s are sold with the excuse that they were "too fast," or "too small," the latter more believable than the former.

Check Title History—Always check the title for information regarding salvage-titles and flood damage through services like Carfax. Keep in mind, though, that a "clean" Carfax doesn't guarantee the car is clean—only that the VIN hasn't had anything reported. Unless you've got experience doing so, and very low expectations, stay away from salvaged-title MR2s. Far more often than finding a bargain, you're going to find trouble.

Check Air Conditioning—How critical this is depends on where you live, but A/C repairs are often very expensive, so when they go, like clutches, instead of repairing them, people often simply sell the car. If A/C isn't critical to you, consider its lack of functionality for future resale: most owners gotta have their A/C.

BUYING A MODIFIED MR2

Many people are averse to buying a modified car, perhaps justifiably so. Having sold several highly modified MR2s over the years, though, to simply not consider them at all would be shortsighted. If you plan on buying one, they can often be had at fire-sale prices, saving you truly impressive amounts of money in the long run, assuming they have been suitably maintained and otherwise responsibly owned.

An MR2 Buyer's Guide

The stock, unrevised MR2 Turbo in the background, or the modified, heavily invested model in the foreground? Their asking price may seem high, but the total investment difference in each car could reach $10,000 or more. Those Volk wheels alone retail for over $2000. Photo courtesy Jensen Lum.

If, on first instinct, you are hesitant to buy a modified car, join the club. The main concern is usually not that the MR2 is too fast but that it may be damaged. No one wants a horse that has been ridden hard and put up wet. There are steps that we can take, though, to at least begin to counter the potential that you're buying damaged goods.

What to Ask

Among those questions I need answered regarding purchasing a modified car:

1. Who, in fact, performed the modifications? If they buyer did, is he qualified to do such installs? Were the installs done well?

2. Do the modifications seem selective and thought-out, or is the car a parts hangar full of non-synchronized junk? Was he thorough, discriminating, and balanced in his approach to performance, or has he been running 19 psi on the stock turbocharger, with worn ignition components and a "custom" exhaust?

3. What is the general quality of the parts installed—are the gauges Sunpro or Blitz?

4. How old are the modifications? Are they up to date? Do they all still work?

5. Why is the current owner selling the car, especially with so much invested? Are they simply moving on to a newer car, or is there more to the story?

Note: On the latter, if you've found the car on a message board, look under the seller's username, under "Search for recent posts," and look through his posts over the years, assuming he's made some. This is a good way to get some additional information on the car, and the seller. If you find a

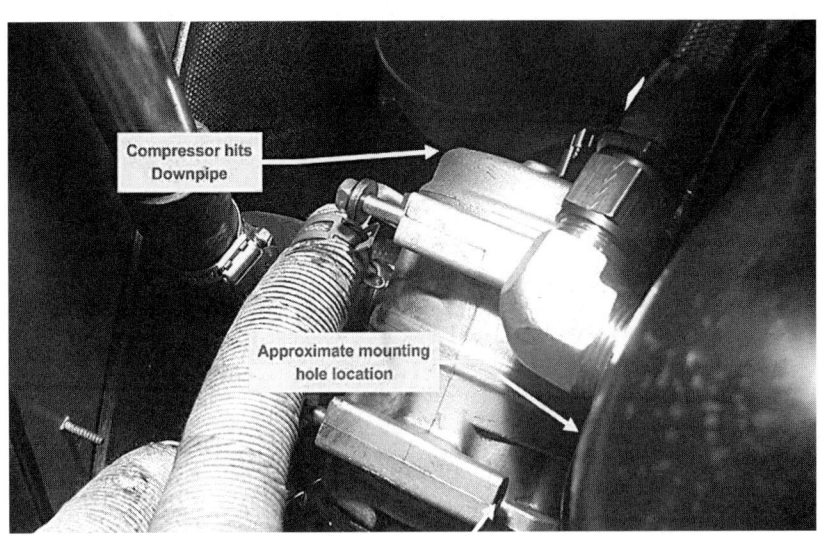

The problem with many aftermarket parts, is fitment—quality. Whether fitment, durability or even, ironically, a lack of performance, the aftermarket isn't always that its is cracked up to be. Here, we can see significant fitment issues with this turbo kit on a 3S-GTE. This is not as uncommon as one would hope.

A boost control and turbo timer, a tachometer mounted on the A-pillar, a flip-screen stereo system, boost and EGT gauges, a performance meter and a radar detector. Thorough, well-chosen upgrades and metering instruments but also that many more things that can go wrong. Photo courtesy Brian Hill.

handful of recent posts discussing the 2010 Subaru WRX he has preordered, you can add some credibility to possibility that the car is sound, and he is simply moving on; if, however, there are a slew of recent posts concerning an intermittent miss or a smoking-at-idle problem—information he conveniently left out of the endorsements he's offered on the car when you have communicated—you at least have reason to keep digging to learn more.

Toyota MR2 Performance

Burning off unspent fuel in the exhaust system when closing the throttle on a dyno looks absolutely spectacular, something a stock SW20 will probably never do. If that makes you more or less likely to purchase such an MR2 used depends on your perspective. Photo courtesy KO Racing.

Stock or modified, hardtop or convertible, AW11, SW2x or ZZW30, the Toyota MR2 is a brilliant runner offering an economically sound sports car ownership experience. Photo courtesy Jim Griffin.

More important than predictors of condition like year, mileage, or even pictures, is the reality of the MR2 sitting in front of you for sale in Anytown, U.S.A. Don't be blinded by low miles, or the latest revision—instead, focus on finding the cleanest, best-kept MR2 you can. Photo courtesy Bryan Heitkotter.

Other posts might tell you a lot about what parts the car has, how old they are, and so on. The seller might not have withheld any of this information from you anyway, but this route can be easier and more direct than asking him a dozen questions in an email. This may sound like snooping, but I call it due diligence. There is a lot of junk being passed around out there.

And be careful—just because something looks amiss doesn't mean that it is. Just like Carfax isn't infallible, neither is this method of info scouring, nor are pictures no matter how detailed they might seem, or any other single source of information. A used car purchase—especially modified, high-performance sports car—must be the product of a synthesis of information and instinct.

A Final Note

When shopping for an MR2, buy the cleanest example you can afford that possesses the coachwork and options that you desire, regardless of year or revision. Generally speaking, the revisions given to MR2s over the years were rarely cause to make previous versions unworthy of your purchase.

The upgrades to the SW20 in 1993 were significant, but the price premium they command isn't absolutely, universally justified by commensurate improvement in the car; remember that it is possible to apply the revisions yourself at any time.

Also, resist the temptation to absolutely, positively have to have buy the lowest-mileage version of the latest revision MR2. While mileage is important, always give the honest reality of the car sitting in front of you the most priority when considering a purchase, above and beyond simple predictors of condition—like mileage.

MR2 RESOURCES

MR2 PERFORMANCE PARTS & SERVICES

ATS Racing
www.atsracing.net
Known For: Their extensive line of SW20 turbocharger kits and available in-shop dyno-tuning
Wrenching Services: Basic MR2 Maintenance, Advanced Performance, Tuning, Dyno
Parts: Proprietary, Major Aftermarket, OEM

AutoLab
www.autolabusa.com
Known For: Excellent fabrication skills and clean engine swaps
Wrenching Services: Basic MR2 Maintenance, Advanced Performance, Tuning, Dyno
Parts: Proprietary, Major Aftermarket

KBox Seals
www.kbox.ca
Known For: Urethane Engine Mounts
Parts: Proprietary, Major Aftermarket, OEM

KO Racing
www.koracing.net
Known For: TKO exhaust and range of excellent turbo kits
Wrenching Services: Basic MR2 maintenance, Advanced Performance
Parts: Proprietary, Major Aftermarket, OEM

MR Controls
www.mrcontrols.com
Known For: Hydra Nemesis Standalone EMS support
Wrenching Services: Advanced Performance Tuning
Parts: Proprietary

MR2Trader.com
Known For: Selling MR2s. The North American distributor for the print magazine MR2Only
Parts: Proprietary, Major Aftermarket

Speed Source
www.speed-source.net
Known For: Custom turbo kit parts, shifter bushings and stainless steel clutch lines
Wrenching Services: Basic MR2 Maintenance, Advanced Performance
Parts: Proprietary, Major Aftermarket

T.E.D. Components
www.tedcomponents.com
Known For: 4A-GE head packages, camshafts upgrades; general cylinder head porting on other engines
Wrenching Services: Advanced Performance
Parts: Proprietary, Major Aftermarket

Twosrus
www.twosrus.com
Known For: Collection of useful MR2 parts, Club MR2
Parts: Proprietary, Major Aftermarket, OEM

WolfKatz Engineering
www.wolfkatz.com
Known For: SW20 fuel system products, an under-developed niche in the big hp market
Parts: Proprietary, Major Aftermarket

The Winning Formula
Known For: Excellent maintenance and performance service for almost any sports car; alignment
Services Offered: Basic MR2 Maintenance, Performance

MR2 CLUBS & ONLINE FORUMS

Clubs

Beyond local city and state-level clubs, there are other MR2 groups that can offer outstanding resources of community and all that community entails, including information exchange, a source of used parts and feedback on major parts from major parts producers like APEXi, HKS, Blitz, Tein and so on.

If you have a Gen III 3S-GTE in your MR2, these clubs, with their forums and email lists, would be particularly useful to you. Many of these countries have been experimenting with Gen III for years, and so have a larger well of experience to draw from. Other countries also have their own way of doing things, and sometimes another way of looking at things is exactly what you and your MR2 need. Those most useful to us as English-speakers will share our language, which rules out significant MR2 markets like those in Japan, Hong Kong and others. However, there are active MR2 clubs in New Zealand, the UK and Australia.

Forum

www.mr2.com

This site offers excellent communities of MR2 owners interested in everything from upgraded stereo systems to all-out racing machines.

MR2 GLOSSARY

The following collected terms represent a quick-reference of commonly used terms and abbreviations for MR2 owners. Some are MR2-specific, while others are standard automotive terms, measures and phrases you need to know anyway.

1500S—Base-model home-market (Japanese) AW10 that featured a 1452cc 3A-LU SOHC engine.

1600G—One step over a 1500S, and one step under a G-Limited, the 1600G featured the venerable 4A-GE engine, but was otherwise a lower-end model.

1600 G-Limited—The premium model AW11 for the Japanese market, no matter its misleading name.

1ZZ-FE—Toyota engine code for the normally aspirated, DOHC, 16-valve, 1.8L MR2 Spyder engine. A version of the 1ZZ-FE was also used in the 2000–2005 Toyota Celica GT.

3S-GE—Toyota engine code for the normally aspirated DOHC, 16-valve, 2.0L engine in international, non-USDM SW21s; also used in other Toyota applications worldwide, such as the Celica and IS200/RS200, and as the foundation for the rightfully celebrated VVT-i and Dual VVT-i BEAMS engines as well.

3S-GTE—Toyota engine code for the turbocharged, DOHC, 16-valve version of the 3S-GE. Used most notably in the SW20 (Gen II), and ST185 (Gen II), ST205 (Gen III) Celica All-Trac applications, and revised again and used in later platforms such as the Toyota Caldina (Gen IV) as well. Most relevant version for USDM MR2 owners would be the Gen II 3S-GTE.
Note: The Celica All-Trac use of the 3S-GTE included use of an air-to-water intercooler.

5S-FE—Toyota engine code for the DOHC, 16-valve, 2.2l USDM SW21 engine. A normally aspirated DOHC, 16-valve, 2.2l engine, it was also used in various Toyota Camry and Celicas throughout the 1990s.

7A-GE—The HKS "5-AG" kit was a well-developed stroker kit for the 4A-GE, but expensive. This "7A-GE" concept was stumbled upon years ago as an answer to cheap stroking. The 7A-FE engine was used during the mid-1990s in Corollas, and features 1.8 liters of displacement, but a poorly flowing cylinder head. Using the larger displacing 7A block and the capable 4A-GE head, a hybrid stroker kit was born. Fittingly enough in true hybrid fashion, you must use a timing belt from Porsche, the 924-944 series, to complete the hybrid. Care must be taken when choosing pistons and associated engine-building components to achieve optimal compression ratios for your application.

AW11—AW11 was the Toyota internal chassis code for the 1985–1989 MR2. "A" designates the 4A-GE engine, "W" designates the MR2 chassis, "11" for the 1985–89 chassis.
Note: In this book, using a mix of chassis codes and VIN identifiers, we've simply used AW15 to refer to all normally aspirated MkIs, AW16 to refer to all supercharged MkIs, and AW11 to refer vaguely to any MkI when the engine designation wasn't necessary.

Area Under the Curve—The area under the curve refers to the part of the graph resting beneath the sweep of the horsepower and torque lines. Graphs that go up at a significant angle, for example, would have less area under the curve than a curve that, while possibly making less peak power, offers a flatter torque curve from just off idle to redline. While area under the curve is important, do not underestimate the usefulness of a transmission, with the best planned gear changes keeping you in the peak portion of the powerband. This isn't always possible though, and is a lot of work, adding tremendous value to an engine able to produce significant power throughout the rev range—one with proper area under the curve.

Bluetop, Redtop—Vague, useless Toyota slang previously used to distinguish between the various forms of 4A-GE engine marked by the color of the lettering on the valve-cover. This method is not specific enough, and a more appropriate way to do so is to refer to the size of the ports, and the number of ribs on the block.

Boost Creep—Boost creep occurs when a wastegate is not able to bleed off enough air to maintain a predetermined boost setting. This is especially problematic with extremely free-flowing exhausts. As rpm increase, boost levels can sky rocket, which sounds fun, but can destroy your engine.

Boost Threshold—The boost threshold of a turbocharger is the rpm at which there is sufficient exhaust gas energy (flow) to spool the turbocharger and create boost pressure. This is different from spool.

Ceramic—A lightweight material with excellent thermal properties used in a variety of automotive applications. In turbocharger technology, the lighter ceramic wheel (as opposed to steel) will offer quicker spooling characteristics. A weakness of such wheels is that they can fail more frequently than steel wheel turbochargers at higher boost levels, generally around 18 psi and above. Obviously a disintegrating ceramic wheels being ingested into the engine would likely cause the engine to fail, something to avoid.

Ceramic CT-26—This is a Japanese Domestic Market factory turbocharger Gen II 3S-GTE, a CT-26 with a ceramic wheel offering quicker spool (see above). The ceramic CT-26 never was offered in a U.S. market MR2, but will indeed bolt right on if you can find one, and offer a decent and acceptable performance gain if you find it cheap and are willing to accept the risk.

Clips—In Japan, cars that have become too expensive to run, for a variety of reasons, are sent to boneyards way before their time, sometimes with as few as 30,000 miles. These cars are then "clipped"—that is they are cut in half so that the still-viable engines, transmissions and other items can be

MR2 Resources

sold. These "clips" are exported all over the world, and have extended the life of many MR2s.

An MR2 clip is generally cut at the rear bulkhead/firewall, behind the seats, and often comes with goodies like taillights, exhaust systems, and other useful items.

CT20B—The stock, twin entry turbocharger on the Japanese Domestic Market Gen III 3S-GTE, and capable of approaching 300 whp with other appropriate modifications, the CT20B is a potent stock turbocharger. It came with both steel and ceramic wheels.

CT26—The stock, twin-entry turbocharger on the Gen II 3S-GTE. This is the turbo the vast majority of MR2 Turbo owners have, and was also used, in slightly different, single-entry form, in the MkIII Toyota Supra Turbo.

CT27—OEM-sounding turbocharger offered by ATS Racing as an upgrade to the CT-26, the CT27 was among the first, and very, very few truly successful, upgrades to the CT26.

BEAMS—Toyota acronym for "Breakthrough Engine with Advanced Mechanism System," used on later 3S-GEs, a BEAMS 3S-GE is an extremely potent, normally aspirated engine.

EBC—Electronic Boost Control. A method of manipulating the wastegate electronically through the use of a stepping motor to increase boost response.

EHPS—Electro-hydraulic Power Steering. It used an electric motor in place of the traditional belt-driven pump that siphoned power from the engine.

G—One of four versions of the Japanese Domestic Market SW2x, the basic "G" grade was the least-optioned MR2, and powered by the 3S-GE engine. Unlike the GT and GT-S, the "G" was available with an automatic transmission, and featured steel rims and fixed foglights.

Gen II 3S-GTE—The second-generation of the 3S-GTE, refined mainly through changes to the cylinder head, turbocharger, and intake manifold. This engine was used in the ST185 Celica All-Trac, and the USDM MR2 Turbo, producing 200 hp in the States, and 225 in JDM dress.

Gen III 3S-GTE—The third-generation of the 3S-GTE, this engine never made it to the United States. The Gen III featured a host of revisions, including the cylinder head, turbocharger, and intake manifold, similar to the Gen II revisions. The Gen III 3S-GTE produced 245 horsepower.

G-Limited—A label not used in the USDM for obvious reasons, like the "G" model the G-Limited MR2 was also normally aspirated, and too was available with an automatic transmission. Rotating foglights were also available on the G-Limited, and it weighed 2596 lbs. in Rev 1 (unrevised) form.
G-Limited Super Edition—A limited edition SW20 released in Japan based upon the G-Limited, its enhancements were mostly cosmetic.

GT—The most expensive JDM MR2, the GT was turbocharged, and heavily optioned, including the aforementioned rotating foglights that rotated in the direction of the turn. The T-top versions of the GT weighed in just under 2800 lbs. in Rev 1 form.

GT-S—While in America GT-S usually means something "extra" over a comparable GT model (see the Toyota Celica), in Japan the GT-S was the less expensive of the two, offering the potent 3S-GTE engine without many of the luxurious amenities the GT held. Both the GT and GT-S were only offered with a 5-speed manual transmission.

Gen I 3S-GTE—The first-generation of the 3S-GTE, a turbocharged version of the 3S-GE. This engine was used in the first generation (ST165) Celica All-Trac, and developed 185 hp.

HKS 5-AG Kit—A stroker crankshaft and oversize pistons developed by HKS to increase the displacement of the 4A-GE to 1720cc. This kit is no longer available new.

Head—The head is basically everything from the valve/cam cover down to the head gasket. On MR2 engines this includes valvetrain, two camshafts, and intake and exhaust ports.

Hose-from-Hell—The hose-from-hell is an aptly named hose on the Gen II 3S-GTE tucked underneath the turbocharger and exhaust manifold that is nearly impossible to get to without pulling all of the turbocharger equipment—thus the troublesome-sounding name. It is also known to fail, requiring consistent (annual or more frequent depending on application) monitoring. This hose was eliminated on the Gen III 3S-GTE.

Lag—Term used to describe the boosting characteristics of a turbocharger. The lag of a turbocharger is the difference between the point where the turbocharger reaches its boost threshold and when it is producing useable boost.

The spool of a turbocharger is observed by how quickly the turbocharger builds boost, most notably the gap between boost threshold and full boost. This is dictated by a number of factors, including the size of the turbocharger relative to the ability of the engine to produce adequate exhaust gas energy, the boost controller, exhaust sizing and general exhaust and wastegate plumbing, and the tune of the air/fuel ratio and engine timing.

Large-Port—4A-GE-speaking, the opposite of "small-port."

Lean—The opposite of rich, where there isn't enough fuel. "Leaning out" a mixture means to remove fuel from a mixture to remedy an overly rich mixture, usually in hopes of generating additional horsepower. While "leaning out" a mixture can make power, it is also dangerous, and should be done so with caution and careful technical consideration.

Longblock—The longblock adds the head to the shortblock, and generally includes manifolds as well and other accessories as well, depending on the application.

Off-the-Shelf—The opposite of one-off, or custom work, off-the-shelf refers to a mass-produced part that is available for purchase by the general public, a term that is usually used to indicate a projects repeatability, or a certain setup's accessibility to the general public.

MBC—Manual boost control. Basically a very simple air-bleed valve that is used to control boost levels.

MkI—Pronounced "Mark One," this collection of letters designates the first generation MR2, introduced in Japan as 1984 model, and in the United States in 1985.

MkII—Pronounced "Mark Two," designating the second generation MR2.

MkIII—Pronounced "Mark Three," designating the third generation MR2 Spyder.

MR2—Originally standing for "Midship, Runabout, 2-seater," in Japan, later U.S. press materials from Toyota explained it to mean "Mid-engine, Rear-drive, 2-seater."

Mandrel-Bent—Mandrel-bent exhaust systems are shaped by a special process that ensures the inside diameter of the piping is not reduced or otherwise compromised by the bend.

Maps (EMS)—Maps are basically numerical values that determine the timing and air/fuel mixture of an engine, as displayed and inputted by the engine management system. Base maps are usually non-optimized maps meant to start the car, and sometimes drive to the dyno or other tuning method to have the car tuned for your specific application.

Plug-and-Play—A term used to describe an item that requires no modification or special install procedure to install and use right away. They are generally easy to install and user-friendly.

Rich—A condition marked by more fuel (or less air) than is ideal to operate an engine. Typically it is safer to be rich than lean, and so stock engines generally run rich.

ST165/ST185/ST205—Chassis code for generations of the turbocharged Celica All-Trac that shared the 3S-GTE engine with the SW20.

SW20, SW21—SW20 was the Toyota internal chassis code for the BEAMS 3S-GE or 3S-GTE MR2 MkII in the Japanese Domestic Market, while the SW21 referred to the 5S-FE MkII and the SW22 referred to the 3S-GTE MkII in the United States Domestic Market. The "S" designates "3S-GTE engine," "W" designates MR2 chassis, "20/21/22" designates the 91–95 generation.

Sherwood—Two-tone metallic paint available on the AW11, the Sherwood paint package featured a beige bottom beneath a dark green "top."

Shortblock—A shortblock is basically the bottom end of your engine—that is, the block, crank, rods, pistons, etc., and may or may not include the oil pan, though it usually refers to everything below the head gasket.

Small Port—A concrete way to distinguish between variations of normally aspirated, 16-valve 4A-GEs, "small-port" indicates the size of the intake ports in the head. Early 4A-GEs featured large ports, but later revisions reduced them in size, combined with higher-compression yielding an increased port velocity and increased performance. The Gen II 3S-GTE also had large ports, while the Gen III version again moved to smaller intake ports.

T-Top—A 2-piece glass-panel removable roof. Also called T-bar roof.

T3/T4—A hybrid turbocharger that has become quite popular, pairing a T3 Turbine section with a T4 compressor.

T78—Another kit from GReddy used for 500+ hp. Laggy even on the larger-displacement 3.0l 2JZ-GTE MkIV Supra, to give you an idea.

TC-16—"Twin Cam, 16-valve," early Toyota corporate speak for the 4A-GE engine.

TD06L2—A variation on the TD06SH, the L2 offered a slightly bigger turbine housing, tubular exhaust manifold, and external wastegate as well. A very nice kit, but pricey.

TD06SH—The 20G turbo kit offered by GReddy for the 3S-GTE, and one of the earliest and most successful aftermarket turbo kit options SW20 owners had. This kit used the stock exhaust manifold. While it offered exceptional performance, it became known for cracked downpipes and boost creep. Considering the higher-quality alternatives on the market, it is no longer the hot setup.

TVIS—Toyota Variable Intake System, basically an intake manifold with vacuum-operated butterfly plates that block off half of the intake runners to increase low rpm performance.

Triangles—MR2-speak for the name of the early stock AW11 wheels; also refers to the "grip-tape" shapes just forward of the rear wheel arches on the lower side-sill of the AW11 exterior.

Twin-Entry Turbocharger—As opposed to a single-entry turbocharger, a twin-entry turbocharger separates the exhaust gas pulses pumping out of the cylinders to keep them from joining unfavorably, reducing lag, and confusing café salesman nationwide.

White Lanner—A Limited Edition AW11 that was all white—bumpers, door guards, side-view mirrors, etc. Only 100 "White Lanner" AW11s were built, and none imported to the U.S.

ZZW30—Toyota internal chassis code for the 2000–2005 MR2 Spyder.

ABOUT THE AUTHOR

Terry purchased his first MR2 in 1996—a black 1992 MR2 turbo. That MR2 won multiple local autocross trophies, was drag-raced during local test-and-tune nights and extensively modified. In addition to reading everything he could get his hands on about these (and other) cars, Terry ventured into the automotive media world, getting his start in online MR2 forums was where he wrote technical answers to questions and learned as much as he could about what others were doing. He followed this path to writing advertising copy and automotive reviews for autowire.net before eventually becoming editor-in-chief at freshalloy.com. Terry also worked with David Vespremi on his book *Car Hacks & Mods For Dummies*.

Terry continued his passion for all things MR2, modifying several normally aspirated AW11 models for hot-lapping road courses, drag racing and autocrossing. For more than a decade he has continued his quest for the perfect "sweet spot" of modification level for each MR2 generation and induction type.

HPBooks

GENERAL MOTORS
Big-Block Chevy Engine Buildups: 978-1-55788-484-8/HP1484
Big-Block Chevy Performance: 978-1-55788-216-5/HP1216
Camaro Performance Handbook: 978-1-55788-057-4/HP1057
Camaro Restoration Handbook ('61–'81): 978-0-89586-375-1/HP758
Chevelle/El Camino Handbook: 978-1-55788-428-2/HP1428
Chevy LS1/LS6 Performance: 978-1-55788-407-7/HP1407
The Classic Chevy Truck Handbook: 978-1-55788-534-0/HP1534
How to Customize Your Chevy Silverado/GMC Sierra Truck, 1996–2006: 978-1-55788-526-5/HP1526
How to Rebuild Big-Block Chevy Engines: 978-0-89586-175-7/HP755
How to Rebuild Big-Block Chevy Engines, 1991–2000: 978-1-55788-550-0/HP1550
How to Rebuild Small-Block Chevy LT-1/LT-4 Engines: 978-1-55788-393-3/HP1393
How to Rebuild Your Small-Block Chevy: 978-1-55788-029-1/HP1029
John Lingenfelter on Modifying Small-Block Chevy Engines: 978-1-55788-238-7/HP1238
Powerglide Transmission Handbook: 978-1-55788-355-1/HP1355
Small-Block Chevy Engine Buildups: 978-1-55788-400-8/HP1400
Turbo Hydra-Matic 350 Handbook: 978-0-89586-051-4/HP511

FORD
Ford Engine Buildups: 978-1-55788-531-9/HP1531
Ford Windsor Small-Block Performance: 978-1-55788-323-0/HP1323
How to Customize Your Ford F-150 Truck, 1997–2008: 978-1-55788-529-6/HP1529
How to Rebuild Big-Block Ford Engines: 978-0-89586-070-5/HP708
How to Rebuild Ford V-8 Engines: 978-0-89586-036-1/HP36
How to Rebuild Small-Block Ford Engines: 978-0-912656-89-2/HP89
Mustang Restoration Handbook: 978-0-89586-402-4/HP029

MOPAR
Big-Block Mopar Performance: 978-1-55788-302-5/HP1302
How to Hot Rod Small-Block Mopar Engine, Revised: 978-1-55788-405-3/HP1405
How to Modify Your Jeep Chassis and Suspension For Off-Road: 978-1-55788-424-4/HP1424
How to Modify Your Mopar Magnum V8: 978-1-55788-473-2/HP1473
How to Rebuild and Modify Chrysler 426 Hemi Engines: 978-1-55788-525-8/HP1525
How to Rebuild Big-Block Mopar Engines: 978-1-55788-190-8/HP1190
How to Rebuild Small-Block Mopar Engines: 978-0-89586-128-5/HP83
How to Rebuild Your Mopar Magnum V8: 978-1-55788-431-5/HP1431
The Mopar Six-Pack Engine Handbook: 978-1-55788-528-9/HP1528
Torqueflite A-727 Transmission Handbook: 978-1-55788-399-5/HP1399

IMPORTS
Baja Bugs & Buggies: 978-0-89586-186-3/HP60
Honda/Acura Engine Performance: 978-1-55788-384-1/HP1384
How to Build Performance Nissan Sport Compacts, 1991–2006: 978-1-55788-541-8/HP1541
How to Build Xtreme Pocket Rockets: 978-1-55788-548-7/HP1548
How to Hot Rod VW Engines: 978-0-91265-603-8/HP034
How to Rebuild Your VW Air-Cooled Engine: 978-0-89586-225-9/HP1225
Mitsubishi & Diamond Star Performance Tuning: 978-1-55788-496-1/HP1496
Porsche 911 Performance: 978-1-55788-489-3/HP1489
Street Rotary: 978-1-55788-549-4/HP1549
Toyota MR2 Performance: 978-155788-553-1/HP1553
Xtreme Honda B-Series Engines: 978-1-55788-552-4/HP1552

HANDBOOKS
Auto Electrical Handbook: 978-0-89586-238-9/HP387
Auto Math Handbook: 978-1-55788-020-8/HP1020
Auto Upholstery & Interiors: 978-1-55788-265-3/HP1265
Custom Auto Wiring & Electrical: 978-1-55788-545-6/HP1545
Engine Builder's Handbook: 978-1-55788-245-5/HP1245
Engine Cooling Systems: 978-1-55788-425-1/HP1425
Fiberglass & Other Composite Materials: 978-1-55788-498-5/HP1498
High Performance Fasteners & Plumbing: 978-1-55788-523-4/HP1523
Metal Fabricator's Handbook: 978-0-89586-870-1/HP709
Paint & Body Handbook: 978-1-55788-082-6/HP1082
Practical Auto & Truck Restoration: 978-155788-547-0/HP1547
Pro Paint & Body: 978-1-55788-394-0/HP1394
Sheet Metal Handbook: 978-0-89586-757-5/HP575
Welder's Handbook, Revised: 978-1-55788-513-5

INDUCTION
Holley 4150 & 4160 Carburetor Handbook: 978-0-89586-047-7/HP473
Holley Carbs, Manifolds & F.I.: 978-1-55788-052-9/HP1052
Rebuild & Powertune Carter/Edelbrock Carburetors: 978-155788-555-5/HP1555
Rochester Carburetors: 978-0-89586-301-0/HP014
Turbochargers: 978-0-89586-135-1/HP49
Street Turbocharging: 978-1-55788-488-6/HP1488
Weber Carburetors: 978-0-89589-377-5/HP774

RACING & CHASSIS
Chassis Engineering: 978-1-55788-055-0/HP1055
4Wheel & Off-Road's Chassis & Suspension: 978-1-55788-406-0/HP1406
Dirt Track Chassis & Suspension: 978-1-55788-511-1/HP1511
How to Make Your Car Handle: 978-1-91265-646-5/HP46
How to Build a Winning Drag Race Chassis & Suspension:
The Race Car Chassis: 978-1-55788-540-1/HP1540
The Racing Engine Builder's Handbook: 978-1-55788-492-3/HP1492
Stock Car Racing Engine Technology: 978-1-55788-506-7/HP1506
Stock Car Setup Secrets: 978-1-55788-401-5/HP1401

STREET RODS
Street Rodder magazine's Chassis & Suspension Handbook: 978-1-55788-346-9/HP1346
Street Rodder's Handbook, Revised: 978-1-55788-409-1/HP1409
Street Rodding Tips & Techniques: 978-1-55788-515-9/HP1515

ORDER YOUR COPY TODAY!

All books are available from online bookstores (www.amazon.com and www.barnesandnoble.com) and auto parts stores (www.summitracing.com or www.jegs.com). Or order direct from HPBooks by calling toll-free at 800-788-6262, ext. 1. Some titles available in downloadable eBook formats at www.penguin.com.